SpringerBriefs in Applied Sciences and Technology

SpringerBriefs present concise summaries of cutting-edge research and practical applications across a wide spectrum of fields. Featuring compact volumes of 50 to 125 pages, the series covers a range of content from professional to academic.

Typical publications can be:

- A timely report of state-of-the art methods
- An introduction to or a manual for the application of mathematical or computer techniques
- A bridge between new research results, as published in journal articles
- A snapshot of a hot or emerging topic
- An in-depth case study
- A presentation of core concepts that students must understand in order to make independent contributions

SpringerBriefs are characterized by fast, global electronic dissemination, standard publishing contracts, standardized manuscript preparation and formatting guidelines, and expedited production schedules.

On the one hand, **SpringerBriefs in Applied Sciences and Technology** are devoted to the publication of fundamentals and applications within the different classical engineering disciplines as well as in interdisciplinary fields that recently emerged between these areas. On the other hand, as the boundary separating fundamental research and applied technology is more and more dissolving, this series is particularly open to trans-disciplinary topics between fundamental science and engineering.

Indexed by EI-Compendex, SCOPUS and Springerlink.

Marco Barla · Alessandra Insana · Sondre Gjengedal · Primož Jelušič · Fleur Loveridge · Iulia Prodan · Grzegorz Ryżyński · Jean de Sauvage · Marta Szlasa · Ana Maria Vieira · Hauke Zachert · Bojan Žlender

Recommendations for the Conceptualisation, Design and Construction of Energy Geostructures

elgip European Large Geotechnical Institutes Platform

Springer

Authors
See next page

ISSN 2191-530X ISSN 2191-5318 (electronic)
SpringerBriefs in Applied Sciences and Technology
ISBN 978-3-032-30926-6 ISBN 978-3-032-30927-3 (eBook)
https://doi.org/10.1007/978-3-032-30927-3

This Springer imprint is published by the registered company Springer Nature Switzerland AG
The registered company address is: Gewerbestrasse 11, 6330 Cham, Switzerland

Marco Barla
Department of Structural, Geotechnical and Building Engineering
Politecnico di Torino
Turin, Italy

Alessandra Insana
Department of Structural, Geotechnical and Building Engineering
Politecnico di Torino
Turin, Italy

Sondre Gjengedal
Norwegian Geotechnical Institute
Oslo, Norway

Primož Jelušič
FGPA
University of Maribor
Maribor, Slovenia

Fleur Loveridge
School of Civil Engineering
University of Leeds
Leeds, UK

Iulia Prodan
Faculty of Civil Engineering
Technical University of Cluj-Napoca
Cluj-Napoca, Romania

Grzegorz Ryżyński
Polish Geological Institute - National Research Institute
Warsaw, Poland

Jean de Sauvage
Université Gustave Eiffel
Bron Cedex, France

Marta Szlasa
Polish Geological Institute - National Research Institute
Warsaw, Poland

Ana Maria Vieira
Geotechnics Department
LNEC
Lisbon, Portugal

Hauke Zachert
Institut für Geotechnik
TU Darmstadt
Darmstadt, Germany

Bojan Žlender
FGPA
University of Maribor
Maribor, Slovenia

Preface

The global demand for energy continues to rise, while the need for renewable energy resources becomes increasingly vital. According to the International Energy Agency, global energy demand is projected to increase by over 25% by 2040, driven by population growth, urbanization, and rising living standards.

Geothermal energy is a renewable energy source that harnesses the heat from within the ground for various purposes and can be classified into two main categories: shallow geothermal energy and deep geothermal energy systems.

Deep geothermal energy systems tap into the higher temperatures found at greater depths below the ground surface. These systems involve drilling deep into the ground to access reservoirs of geothermal fluids. The heat extracted from the deep geothermal reservoirs can be used for both electricity generation and direct heating applications.

Shallow geothermal energy (SGE), often accessed by ground-source heat pump (GSHP) systems, utilizes the relatively constant temperature of the upper meters of the ground surface. The principle behind GSHP is to transfer heat from the ground into a building for heating purposes in winter and can also provide cooling in reverse mode during summer season. Shallow geothermal systems are commonly used for space heating, cooling, and domestic hot water heating in residential, commercial, and institutional buildings.

In addition to the classic closed-circuit systems (geothermal probes) and open circuits (geothermal wells), in recent decades the concept of energy geotechnical structures (or simply energy geostructures) has spread. Energy geostructures represent an innovation in shallow geothermal systems. In these systems, the heat exchange circuit is incorporated into structural elements mainly in contact with the ground, with obvious cost advantages with respect to traditional closed circuit geothermal installations.

By incorporating energy geostructures into building design and construction, we can tap into the vast potential of renewable energy resources, contribute to mitigating climate change, and pave the way towards a more sustainable and energy-efficient future. Consequently, every new construction in contact with the ground that does not include for heat exchangers to become an energy geostructure can be viewed as a missed opportunity.

Different experiences have recently grown in Europe leading to the European Commission to fund the COST Action TU1405 GABI (European network for shallow geothermal energy applications in building and infrastructures) that allowed, between 2015 and 2019, networking and sharing among European experts in the field of energy geostructures. This led to the recognition of an opportunity to begin the development of recommendations dedicated to the design and construction of energy geostructures at a European level. The resulting document was finalized by the European Large Geotechnical Institute Platform (ELGIP) and its Energy geostructures Working Group in collaboration with the following COST Action CA21156 FOLIAGE (Fostering large-scale implementation of energy geostructure), active between 2023 and 2026.

These recommendations illustrate best practice and state of the art for the design, construction and monitoring of energy geostructures with the ambitious goal of representing a practical and accessible guide for all public and private actors, professionals in the construction and geotechnical world and those who are interested in learning more about energy geostructures. The recommendations have been written with the intention of providing a tool for organic and integrated understanding of the design process of energy geostructures which allows for the best possible approach to studies, conceptual modelling, surveys, projects, monitoring, short- and long-term modelling, predictions and scenarios of environmental impacts.

Designing energy geostructures for small and large projects involves several key steps to ensure their successful implementation: investigation, design, execution and monitoring. When designing energy geostructures, the complexity of the project, size, and site characteristics play a crucial role in determining the design approach and level of detail for analysis. While small projects may not require advanced and complex numerical approaches for proper functioning during exploitation, the success and well-functioning of big projects is directly linked to the complexity and details considered in the design.

It should be noted here that at present there are some reference documents in several countries (Canada, France, Italy, Switzerland, UK and USA) and the second generation of Eurocode 7 will have an Informative Annex with general indications on energy geostructures.

The publication of this document was supported by the European Large Geotechnical Institutes Platform (ELGIP) and the COST Action CA21156 FOLIAGE (Fostering large-scale implementation of energy geostructure).

The document is structured as follows, making reference to the various design phases (feasibility study, outlined design, final design).

Chapter 1 provides a background to the use of energy geostructures in the context of sustainability.

Chapter 2 introduces the additional investigations and design data to take into account for the design of energy geostructures.

Chapter 3 illustrates the tools currently available to proceed with the thermal design of thermally activated piles, retaining structures, slabs and tunnels. These tools are divided into methodologies aimed at a rapid preliminary evaluation and analytical, semi-analytical and numerical calculation methods, aimed at achieving a greater degree of detail.

Chapter 4 deals with the topic of geotechnical design, clarifying calculation techniques available to evaluate the effects of the variations of temperature in the structures.

Chapter 5 is dedicated to the environmental aspects including thermal interferences and interaction among geothermal systems, while Chap. 6 is dedicated to provide some indications for the execution and monitoring stages.

Three selected examples with their relevant data are included in Appendix.

Turin, Italy — Marco Barla
Turin, Italy — Alessandra Insana
Oslo, Norway — Sondre Gjengedal
Maribor, Slovenia — Primož Jelušič
Leeds, UK — Fleur Loveridge
Cluj-Napoca, Romania — Iulia Prodan
Warsaw, Poland — Grzegorz Ryżyński
Bron Cedex, France — Jean de Sauvage
Warsaw, Poland — Marta Szlasa
Lisbon, Portugal — Ana Maria Vieira
Darmstadt, Germany — Hauke Zachert
Maribor, Slovenia — Bojan Žlender

Acknowledgements

This publication is based upon work from COST Action Foliage CA21156, supported by COST (European Cooperation in Science and Technology).

COST (European Cooperation in Science and Technology) is a funding agency for research and innovation networks. Our Actions help connect research initiatives across Europe and enable scientist to grow their ideas by sharing them with their peers. This boosts their research, career and innovation.

www.cost.eu

Contents

Chapter 1
Contribution of Energy Geostructures to Sustainable Development

Abstract This chapter frames energy geostructures within the global transition to sustainable energy systems. It highlights the environmental impact of fossil fuel consumption and the urgent need for renewable, low-carbon solutions, particularly in the building sector, which accounts for a large share of energy use. Energy Geostructures (EGs) are presented as an innovative form of shallow geothermal energy that integrates heat exchangers into structural elements such as piles, walls, and tunnel linings. This dual functionality reduces costs and environmental impact while improving efficiency. The chapter emphasizes the role of EGs in mitigating climate change, reducing CO_2 emissions, and supporting energy-efficient urban development.

The rise in global population and the increasing use of energy primarily from fossil fuels have had, and continue to have, a significant impact on the environment. This is clearly seen in climate change resulting from substantial greenhouse gas emissions. In light of these concerning trends, transitioning to renewable and local energy sources has become more critical than ever. Moreover, beyond the immediate environmental impacts, the finite nature of fossil fuels and their unsustainable usage further compounds the urgency of this transition.

The shift towards the use of renewable energy sources aims to lessen our contribution to global warming and involves a two-fold approach. The first is mitigation, which focuses on reducing CO_2 emissions and directly tackling the underlying cause of global warming. The second is adaptation, dealing with the consequences of climate change, including addressing heat islands that disproportionately affect urban areas.

European energy consumption continues to rise each year, with buildings accounting for more than 30% of the total energy usage (REN21 2022). Within this building energy consumption, space heating, cooling in residential areas, and hot water collectively contribute to as much as 74% of the consumption (REN21 2022). Given the urgency of dealing with the climate change crisis, the development of local and cost-effective low-carbon energy sources with minimal environmental impact is of paramount importance.

M. Barla et al., *Recommendations for the Conceptualisation, Design and Construction of Energy Geostructures*, SpringerBriefs in Applied Sciences and Technology,
https://doi.org/10.1007/978-3-032-30927-3_1

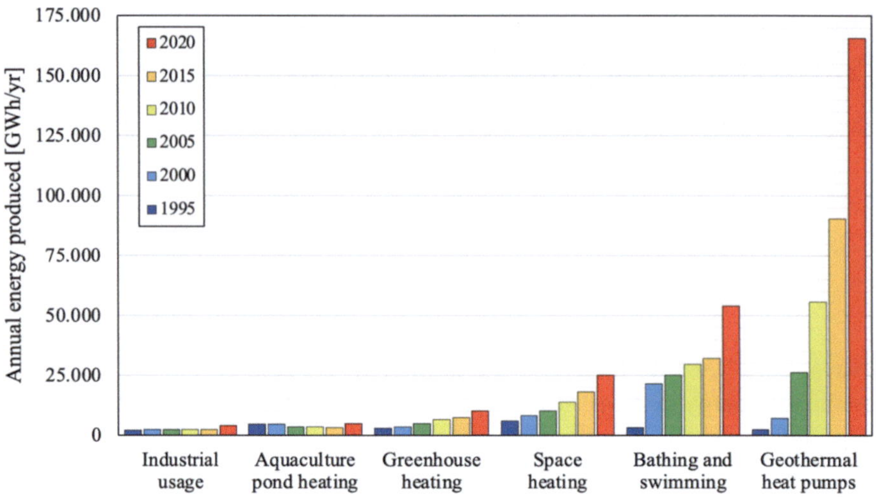

Fig. 1.1 Comparison of worldwide direct-use of geothermal energy in GWh/yr from 1995, 2000, 2005, 2010, 2015 and 2020 (Lund and Toth 2020)

The carbon intensity of electricity has consistently decreased since 1990 in many European countries. Within the European Union, the carbon intensity of electricity used at low voltage dropped from 641 $gCO2_{eq}$/kWh in 1990 to 334 $gCO2_{eq}$/kWh in 2019 (Scarlat et al. 2022). The declining carbon intensity of electricity underscores the need to enhance the use of renewable energy sources. As this trend continues, there is a vital push towards accelerating the transition to sustainable energy solutions.

Additionally, as electricity decarbonises, all renewable energy options connected to heat pumps are increasingly environmentally friendly. This highlights the importance of promoting and expanding the adoption of heat pumps, considering their positive contributions to addressing climate concerns.

SGE offers distinct advantages, including its widespread adaptability and local availability. Notably, GSHPs outperform air-air heat pumps in terms of the Coefficient of Performance (COP), as evidenced by Staffell et al. (2012). This performance discrepancy arises from the greater temperature gradient achievable between the energy source and the user. Moreover, GSHPs circumvent the challenge of releasing excess heat into urban environments that are already overheated.

For a comprehensive understanding of current installation trends, up to date installation information is given in Fig. 1.1 (Lund and Toth 2020).

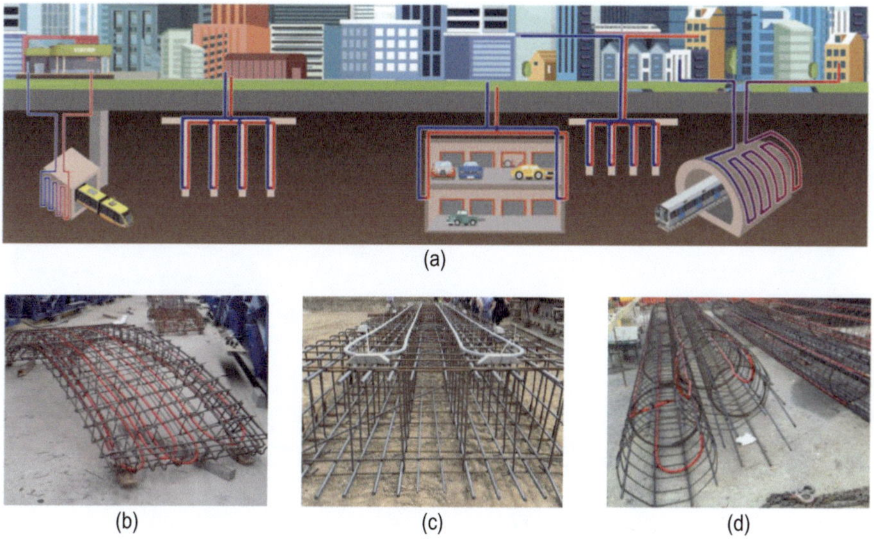

Fig. 1.2 **a** Illustration of the dual role of energy geostructures (courtesy of Foliage Cost Action) and example photos of, **b** energy tunnel lining (Politecnico di Torino, Italy), **c** energy diaphragm wall (Politecnico di Torino, Italy) and **d** energy pile (courtesy of Eneren, Italy)

Geothermal energy systems, despite their vast potential, appear to be undervalued in terms of implementation numbers when compared to other renewable sources. A significant hurdle to their widespread adoption is the initial installation cost. This could be significantly decreased by the adoption of energy geostructures, considering the cost savings due to the reduced need for specific drilling and/or excavation.

Since the 1980s, the development of Energy Geostructures (EGs) has revolutionized the utilization of SGE. This involves harnessing the thermal potential of structural elements in contact with the ground, such as pile foundations, retaining walls, slabs, and tunnels, by incorporating heat exchanger pipes within them (Fig. 1.2). Essentially, these pipes can be affixed to the reinforcement cages of reinforced concrete piles or diaphragm walls, facilitating heat exchange with the building.

EGs represent a groundbreaking advancement in geothermal energy technology, as they repurpose structural components for a dual role as heat exchangers. This innovative approach effectively combines structural integrity with energy harvesting capability. This dual role, combining structural and energy harvesting functions, yields great sustainability, as the concrete elements are already necessary for reasons of resistance and stability.

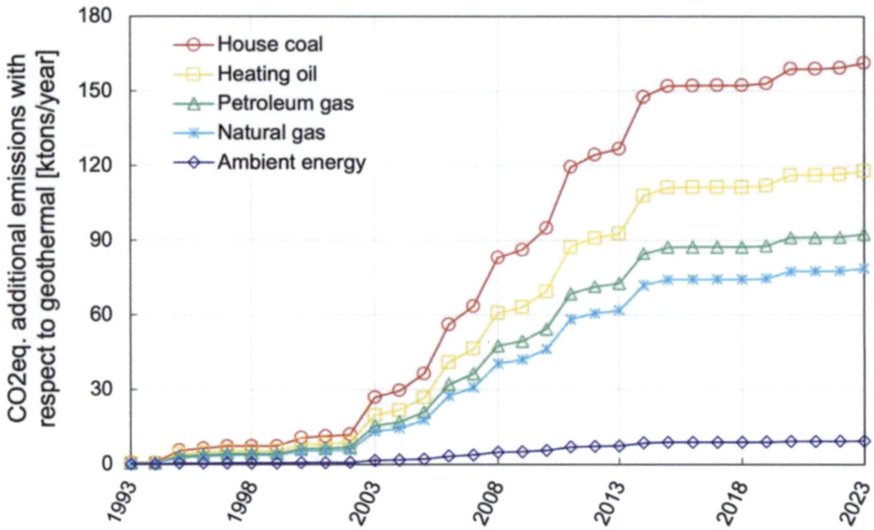

Fig. 1.3 CO_2 equivalent emissions for thermal energy production from fuels (house coal, heating oil, petroleum and natural gases) or air source heat pumps with respect to geothermal energy (Salciarini et al. 2025)

Moreover, EGs are closed-loop geothermal solutions, limiting the environmental impacts sometimes associated with open-loop systems that involve induced groundwater fluxes. Their implementation enables the harnessing of local energy, reducing the demand for scarce green electricity, and they can be easily integrated with other renewable energy sources.

EGs have been under development for the last decades (Brandl 2006), and their global installations are on the rise. Di Donna et al. (2017a) and Salciarini et al. (2026) quantified the reduction in CO_2 emissions attributed to energy geostructures, and Fig. 1.3 illustrates these cumulative annual CO_2 savings since 1993, highlighting the significant contribution of EGs.

Assessing economic benefits poses challenges due to data limitations; however, available estimated payback periods for EGs typically fall between 4 and 8 years, as seen in the case presented by Poulsen et al. (2022). This range is notably influenced by site-specific conditions and the prevailing dynamics of the energy market.

Since Borehole Heat Exchangers (BHEs) and Energy Piles (EPs) are highly similar technologies, it is reasonable to compare their respective costs and benefits. This question becomes particularly critical in sites where the spatial constraints are significant (and these sites generally demand substantial energy). In this context, Di Donna et al. (2017a) conducted a comparison of the relative costs between BHEs and EPs, which can be found in Table 1.1. The primary distinction lies in the utilization of existing piles for EPs, whereas BHEs necessitate supplementary construction, resulting in added costs, extended timelines, and increased environmental impact.

Table 1.1 Comparison between cost benefit of BHEs and EPs (modified from Di Donna et al. 2017a)

Item	Borehole loop	Thermal pile
Diameter (m)	0.2	0.6
Heat transfer (W/m)	35	35
Number of loops	One	Two
Length	100 m (3500 W)	27 m × 2 No (1750 W)
Boring/installation cost	\$50/m × 100 = \$5000	Pile included anyway Allow 2 h′ attach loops \$270/crew time
Thermal grout	\$7/m × 100 m = \$700	N/A
Pipe—25 mm ID	\$7/m—1 loop = \$700	\$7 × 2 × (27 m × 2) = \$756
Spoil removal	\$1000	Nil
Total installation	\$7400	\$1026
Header pipes	\$1500/borehole + trenching	\$1100 per pile
Grand total for 3500 W	\$8900 + trenching	\$3226 (2 piles)
Construction work	Additional construction time required if under footprint of building	On critical path. Addition of loops in piles adds no additional construction period

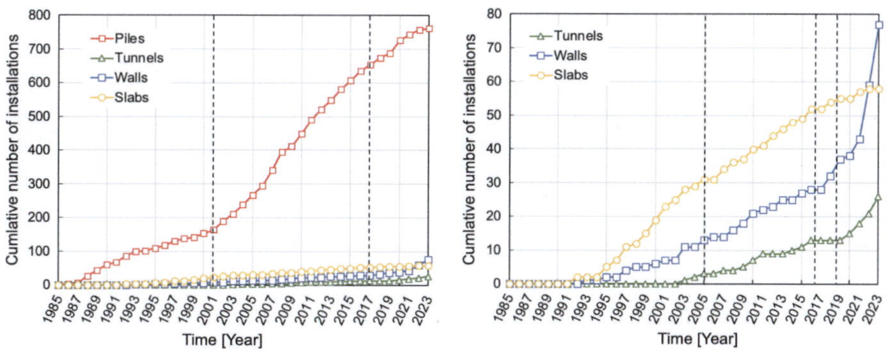

Fig. 1.4 Evolution of EGs installations (Salciarini et al. 2026)

Additionally, the confluence of environmental and economic benefits associated to EGs has been a driving force in their advancement since the 1990s (as depicted in Fig. 1.4).

Chapter 2
Design Data: Site Characterisation and Thermal Loading Conditions

Abstract This chapter outlines the essential data required for designing energy geostructures, focusing on ground characterization and thermal loading. It distinguishes between feasibility-stage assessments and detailed investigations, covering thermal, hydrogeological, and thermomechanical properties of soils. Key parameters include ground temperature, thermal conductivity, heat capacity, and groundwater conditions. The chapter also discusses methods such as thermal response tests and laboratory analyses. Thermal loading is treated as a time-dependent action influenced by building energy demand and seasonal variations, often simplified through spatial and temporal assumptions. Accurate site characterization is emphasized as critical for system performance, efficiency, and long-term sustainability.

The design of Energy Geostructures (EGs) involves both energy and geotechnical considerations. It is essential to have an adequate characterization of the ground for both aspects, as having a proper understanding of ground conditions for EG systems aims to determine:

- whether the ground can meet the energy demand of the intended structure or infrastructure (e.g., Brandl 2006; Barla and Di Donna 2018; Loveridge et al. 2020);
- if the structural and geotechnical integrity of the geostructure is influenced by the additional stresses and strains arising from cyclic temperature fluctuations (Laloui et al. 2014; Rotta Loria and Laloui 2017; Insana 2020; Insana et al. 2020);
- the possible development of thermal plumes in the ground, which could impact the energy efficiency of the system in the long-term, by investigating the soil thermal properties and the hydrogeological setting.

The level of detail of the site characterization for EGs depends on the design stage and on the relevance of the system itself. As stated in EN 1997-2, preliminary investigations should enable the engineer to evaluate the overall suitability of the site, estimate the potential impacts of the proposed works, and effectively plan the follow-up design and control investigations. In general, at least two stages can be considered for site characterization:

M. Barla et al., *Recommendations for the Conceptualisation, Design and Construction of Energy Geostructures*, SpringerBriefs in Applied Sciences and Technology,
https://doi.org/10.1007/978-3-032-30927-3_2

- **Feasibility design**: a preliminary stage for energy design which occurs at the planning phase when the conceptual idea of the project is detailed, where the geological and hydrogeological conditions are investigated and a first estimate of thermal parameters is provided to assess the system's feasibility based on the expected building or infrastructure energy demand;
- **Outlined and detailed characterisation**: a detailed characterisation stage where the final ground design parameters are established, the energy design is reevaluated and the geotechnical design is performed.

2.1 Feasibility Design

During the feasibility or planning phase of an EGs, it is essential to conduct an initial assessment of the on-site conditions and the thermal characteristics of the soil. The most significant parameters for the characterization of the ground thermal behaviour for EGs include (Vieira et al. 2022):

- the undisturbed ground temperature, denoted as T_0 [K], represents the average temperature of the surrounding ground prior to the operation of the EG system;
- the ground thermal conductivity, refers to the soil's ability to conduct heat. It stands as the most critical parameter for assessing the energy performance of the EG. Soils with higher thermal conductivity facilitate greater heat transfer rates and enable faster recovery from thermal depletions and build-ups;
- the ground thermal resistance, is a measurement of the ground-structure resistance to heat flow;
- the ground volumetric heat capacity is the capacity of the ground to store heat. It usually ranges from 1 to 3 $MJ/(m^3 \cdot K)$;
- the depth of the phreatic surface and the characteristics of groundwater flow.

At this initial stage, these parameters can be estimated based on geological mapping, published values, case studies and expert judgement. More recently maps of shallow geothermal potential have been developed locally for undisturbed or thermally affected conditions, such as in dense urban areas (Baralis and Barla 2024).

Undisturbed ground temperature can be evaluated as the average temperature at the site over the depth of the EG. For the preliminary design, the annual average atmospheric temperature can be considered. More precision can be obtained by in situ measurements.

Thermal conductivity and heat capacity depend on different factors, with the most important being the water content, the porosity and the mineralogical composition. Thermal conductivity also depends on spatial arrangement of the different soil phases. These parameters will vary with depth and different approaches can be used to estimate average values. A preliminary estimate can be obtained by consulting regional maps which can be found online and in the literature for several regions (e.g. Ditlefsen et al. 2014; Ramstad et al. 2015; Geothermische Screeningstool, https://tool.smartgeotherm.be/geo/alg). An example is the Thermomap project that provided 10-m average thermal conductivity values for Europe (Fig. 2.1), based on an exhaustive compilation of available data. However, due to their scale, these maps do not incorporate local site effects or specific conditions, which can be particularly relevant at the project scale. Complementary information to estimate the ground thermal characteristics can be derived from published data on thermal properties of soils, provided that the geological setting of the considered area is known. Ranges of values of thermal conductivity and volumetric heat capacity of soils and other materials can be found in the literature (e.g. Farouki 1981; VDI 2010). Some representative values are presented in Fig. 2.2. Moreover, a recent study by Dalla Santa et al. (2020) provides an extensive compilation of thermal conductivity test results, which constitutes a relevant database for studying this property in rocks and unconsolidated sediments. This database was created as part of a decision support system project for the use of GSHP systems.

Thermal resistance can be simply expressed as the ratio of the temperature difference between the heat carrier fluid and the ground and the applied heat flux. It depends on the number of pipes and their geometric arrangement and the physical properties of the EGs elements. A key factor in this regard is the thermal conductivity of the material between the heat exchanger pipes and the surrounding ground. A lower thermal resistance implies more efficient heat transfer within the EG, contributing to overall system efficiency. As for borehole heat exchangers (BHEs), there are several publications in the literature providing calculation formulas for the thermal resistance of different EGs, such as energy piles (Loveridge and Powrie 2013a; Claesson and Javed 2020), tunnel linings (Shafagh et al. 2020) and diaphragm walls (Shafagh and Rees 2019; Gupta 2024).

The position of the water table is crucial, as it determines the thermal properties of the soil, such as thermal conductivity, and thus the efficiency of the system. Additionally, groundwater flow can provide significant additional heat transfer by advection

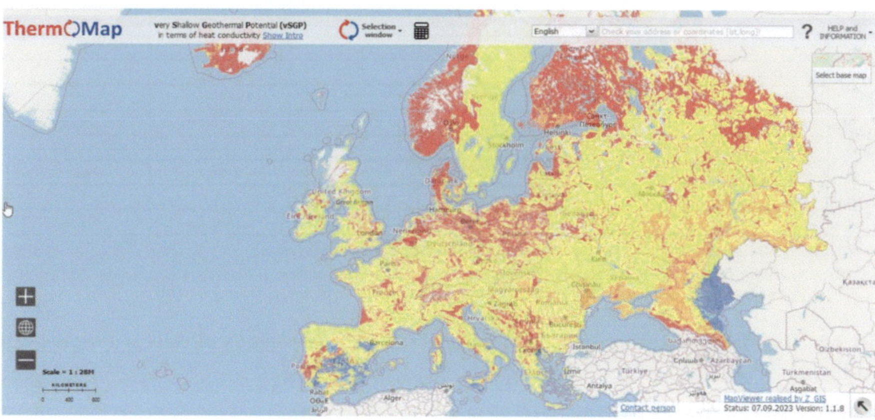

Fig. 2.1 Very shallow geothermal potential in terms of heat conductivity of unconsolidated underground up to 10 m depth (Vieira et al. 2022)

	Type of rock		Thermal conductivity λ in W/(m·K)		Volume-related specific heat capacity $\rho \cdot c_p$ in MJ/(m^3·K)	Density ρ in 10^3 kg/m^3
				recommended value		
Unconsolidated	clay/silt, dry		0,4–1,0	0,5	1,5–1,6	1,8–2,0
	clay/silt, water-saturated		1,1–3,1	1,8	2,0–2,8	2,0–2,2
	sand, dry		0,3–0,9	0,4	1,3–1,6	1,8–2,2
	sand, moist		1,0–1,9	1,4	1,6–2,2	1,9–2,2
	sand, water-saturated		2,0–3,0	2,4	2,2–2,8	1,9–2,3
	gravel/stones, dry		0,4–0,9	0,4	1,3–1,6	1,8–2,2
	gravel/stones, water-saturated		1,6–2,5	1,8	2,2–2,6	1,9–2,3
	till/loam		1,1–2,9	2,4	1,5–2,5	1,8–2,3
	peat, soft lignite		0,2–0,7	0,4	0,5–3,8	0,5–1,1
Sedimentary rock	clay/silt stone		1,1–3,4	2,2	2,1–2,4	2,4–2,6
	sandstone		1,9–4,6	2,8	1,8–2,6	2,2–2,7
	conglomerate/breccia		1,3–5,1	2,3	1,8–2,6	2,2–2,7
	marlstone		1,8–2,9	2,3	2,2–2,3	2,3–2,6
	limestone		2,0–3,9	2,7	2,1–2,4	2,4–2,7
	dolomitic rock		3,0–5,0	3,5	2,1–2,4	2,4–2,7
	sulphate rock (anhydrite)		1,5–7,7	4,1	2,0	2,8–3,0
	sulphate rock (gypsum)		1,3–2,8	1,6	2,0	2,2–2,4
	chloride rock (rock salt, potash)		3,6–6,1	5,4	1,2	2,1–2,2
	anthracite		0,3–0,6	0,4	1,3–1,8	1,3–1,6
Magmatic rock	tuff		1,1	1,1		
	vulcanite, acid to intermediate	e.g. rhyolite, trachyte	3,1–3,4	3,3	2,1	2,6
		e.g. latite, dacite	2,0–2,9	2,6	2,9	2,9–3,0
	vulcanite, alkaline to ultra-alkaline	e.g. andesite, basalt	1,3–2,3	1,7	2,3–2,6	2,6–3,2
	plutonite, acid to intermediate	granite	2,1–4,1	3,2	2,1–3,0	2,4–3,0
		syenite	1,7–3,5	2,6	2,4	2,5–3,0
	plutonite, alkaline to ultra-alkaline	diorite	2,0–2,9	2,5	2,9	2,9–3,0
		gabbro	1,7–2,9	2,0	2,6	2,8–3,1
Metamorphic rock	slightly metamorphic	clay shale	1,5–2,6	2,1	2,2–2,5	2,4–2,7
		chert	4,5–5,0	4,5	2,2	2,5–2,7
	moderately to highly metamorphic	marble	2,1–3,1	2,5	2,0	2,5–2,8
		quartzite	5,0–6,0	5,5	2,1	2,5–2,7
		mica schist	1,5–3,1	2,2	2,2–2,4	2,4–2,7
		gneiss	1,9–4,0	2,9	1,8–2,4	2,4–2,7
		amphibolite	2,1–3,6	2,9	2,0–2,3	2,6–2,9
Other materials	bentonite		0,5–0,8	0,6	~3,9	
	concrete		0,9–2,0	1,6	~1,8	~2,0
	ice (–10 °C)		2,32		1,87	0,919
	synthetics (HD-PE)		0,42		1,8	0,96
	air (0 °C to 20 °C)		0,02		0,0012	0,0012
	steel		60		3,12	7,8
	water (+10 °C)		0,59		4,15	0,999

Note 1: The density of unconsolidated rock varies greatly with consolidation and water content.

Note 2: In sandstone, conglomerates and breccia, there is an especially wide range of thermal conductivities. The important factors, in addition to the constituting material and its distribution and water saturation, are the type of binding material or matrix.

Fig. 2.2 Thermal conductivity and volume-related specific heat capacity of soils and rocks (VDI 2010) in SGE systems (Di Donna and Barla 2016; Di Donna et al. 2021; Claesson and Hellström 2000). Its impact on geothermal system performance differs, depending also on the ground use, i.e., on the thermal needs of the building. The *groundwater flow* can be beneficial for ground source systems that do not rely on seasonal storage, since it would recharge the heat faster or would remove the heat away (Laloui and François 2009; Ma and Grabe 2010; Suryatriyastuti 2013; Loveridge et al. 2017; Epting et al. 2020; Insana and Barla 2020). According to SIA (2005), the seasonal heat storage becomes unfeasible when Darcy's velocity exceeds 0.5–1 m/day. Hence, the geological and/or hydrogeological surveys of each region should be consulted to obtain meaningful information.

2.2 Outlined and Detailed Characterisation

Soil thermal properties and hydrogeological conditions are crucial for the proper dimensioning of SGE systems, as they directly influence the system's long-term energy efficiency (Vieira et al. 2022). These factors are also key in the creation of thermal plumes, which can form a permanent thermal zone affecting the system's performance. Reliable project input data (parameters), based on thorough site investigation, is essential to ensure the good operation and performance of SGE systems. Blum et al. (2011) analysed cases (more than 1000 SGE systems installed in Germany) focusing on the consequences of an inadequate determination of site characteristics to the performance of the SGE system. They concluded that successful planning of the operation of these systems is directly related to sufficient site investigation and on data reliability. This lesson applies directly to EGs and therefore such detailed thermal characterization is crucial. However, the evaluation of soil thermal properties is generally not considered in routine site investigations, and additional characterization studies are required.

Finally, the functionality and safety of EGs in the context of structural and geotechnical design depends on the thermo-hydro-mechanical behaviour of the geostructure and soil-structure interaction, focusing on the relative difference in thermal expansion and support conditions of the geostructure.

2.2.1 Detailed Thermal Characterization

The final design of the EG should be based on undisturbed ground temperature, preferably determined with in-situ tests. A comprehensive review of methods for determining undisturbed ground temperature can be found in Vieira et al. (2017). Measurements can be made using either the borehole temperature measurement or the fluid circulation method from thermal response tests (see below). In borehole temperature measurement, the temperature distribution along the borehole depth can be measured using a downhole temperature measurement system. A simple or weighted average of the measured temperature values can then be used to approximate the undisturbed ground temperature. A variety of systems can be used to measure downhole temperature, including wired temperature sensors, wireless submersible probes, and fibre-optic systems. In other ways, the undisturbed ground temperature can be determined by circulating a fluid through a closed loop ground heat exchanger (GHE) without heating or cooling the fluid and measuring the fluid outlet temperature from the GHE.

Ground thermal properties (thermal conductivity and specific heat) can be determined by in-situ tests or by laboratory tests. They can also be estimated by semi-empirical models based on the relative proportions of soil phases (solid, liquid and gas). Several expressions can be found in the literature that assume different distributions and geometric arrangements (Dong et al. 2015), with parallel and serial configurations providing lower and upper bounds on thermal conductivity, respectively. According to Rees et al. (2000), the weighted geometric mean configuration was found to be suitable for a variety of soils by several researchers. According to them, the thermal conductivity of the soil is:

$$\lambda = \lambda_s^{\chi_s} \cdot \lambda_w^{\chi_w} \cdot \lambda_a^{\chi_a} \tag{2.1}$$

where λ_s, λ_w and λ_a are the thermal conductivities of solid, water, and air, respectively, and $\chi_s = 1 - n$, $\chi_w = n \cdot S_r$ and $\chi_a = n \cdot (1 - S_r)$, are the respective volume fractions, n is the porosity, and S_r is the saturation ratio.

Since the heat capacity of soils is not a directional variable, it can be determined by summing the heat capacities of the various constituents based on their volume fractions, using:

$$c_p = \chi_s \cdot c_s + \chi_w \cdot c_w + \chi_a \cdot c_a \tag{2.2}$$

where c_s, c_w and c_a are the heat capacity of the solid phase, the water phase and the air phase, respectively.

The Thermal Response Test (TRT) (Gehlin 2002) is an in-situ test conducted in an already constructed GHE to determine the thermal conductivity of the surrounding soil, the thermal resistance of the GHE, and the undisturbed initial temperature. In a TRT, heat is injected into the ground at a constant rate in a BHE and the temperature change of the circulating fluid is monitored. Because of the simplicity of design, control, and implementation, this is the preferred method for estimating thermal properties. Originally developed for vertical BHEs, TRT (Spitler and Gehlin 2015) has been extended to energy piles (GSHPA 2012; Jensen-Page et al. 2019; IGSHPA 2025). For energy piles, the principle of the test remains the same as for BHEs, but longer times are usually required to overcome the thermal inertia of the concrete pile. The most up to date guidance is provided in the 2025 CSA/ANSI standard C448 SERIES. This recommends some typical test times for overcoming the thermal capacity of the concrete for different pile sizes. For example, a 2.5-day TRT for 300 mm diameter piles, or a test in excess of 10 days for 600 mm diameter piles. For cases where a shorter test is desirable, then two options are suggested, either to adopt alternative interpretation methods with a transient component for the pile concrete, e.g. Loveridge and Powrie (2013a), Loveridge et al. (2022), Maragna and Loveridge (2019), or to test a small diameter borehole of the same length.

The ground thermal conductivity and the thermal resistance of the GHE, determined by the TRT, have opposite effects on the design. Using the experimentally determined values of both parameters reduces some of the errors that would occur if only the value of ground thermal conductivity were estimated from the tests used for design (Javed et al. 2011). However, it is recommended that the thermal resistance value of the GHE be calculated separately to control the experimentally determined value. It is recommended that the thermal resistance be determined using the Multipole method (Claesson and Hellström 2011). This is an analytical method that can be applied to any number of arbitrarily arranged pipes in a composite area with remarkable accuracy.

TRT has also been used to estimate the ground volumetric heat capacity. Wagner and Clauser (2005) used synthetic TRT data, to make a parameter estimate varying the thermal conductivity and volumetric heat capacity of soil in a finite difference model. In addition, Christodoulides et al. (2016) proposed a method to calculate the ground thermal diffusivity, volumetric heat capacity, and thermal resistance of a GHE. The method was based on actual parameters used in an experiment from TRT and the line source model. However, Austin et al. (2000) points out that the volumetric heat capacity values estimated by TRT are not always acceptable. Therefore, it is recommended that a value for ground volumetric heat capacity be assigned based on knowledge of the soil present and treated as a known value. This assumption will not have a significant impact on the final estimate of the thermal conductivity of the soil. Although less common because they are less easy to perform, in situ tests can also be used to determine the specific heat capacity of soils by constructing pits approximately 1.5 m below the ground surface in and around the SGE plant (Oladunjoye and Sanuade 2012).

It is difficult to determine the ground thermal conductivity and the GHE thermal resistance from TRT test on EGs that are not energy piles, and this is therefore not a common practice. However, some authors perform thermal performance tests to determine the efficiency of EGs, e.g., for energy slabs (Lee et al. 2021), energy walls (Di Donna et al. 2017b), and tunnel linings (Barla et al. 2019). Note that these tests do not provide information about the ground properties. In addition, for energy tunnels and diaphragm walls, the climate within the excavation plays an important role in heat transfer processes that can critically affect thermal performance (Bourne-Webb et al. 2016a; Dornberger et al. 2022; Ma et al. 2021). Therefore, a good understanding of the expected airflow temperature and velocity is also recommended.

The results of TRT measurements can be affected if a sufficiently high groundwater flow occurs. The increased heat transfer due to groundwater advection results in a higher effective thermal conductivity of the ground. The advantage of TRT is that it yields an effective ground conductivity that already accounts for the effect of groundwater flow. However, to decouple conduction and advection effects, it is recommended that a stepwise analysis be performed for this type of data or that a moving line source solution be used (Diao et al. 2004; Sanner et al. 2007).

Laboratory tests can also be used to characterize the thermal parameters of the ground. Tests may be steady state or transient, some of which are standardized, although not always for soil and rock applications. The main advantages of laboratory tests are that they are fast, relatively inexpensive, and allow better control of boundary conditions. However, they do not consider real field conditions, and due to the heterogeneity and variability of the ground many more measurements are required to obtain representative values of thermal properties (King et al. 2012). Steady state tests (such as the divided bar for rocks) are theoretically more reliable, but in soils the long duration of tests may lead to moisture content changes, meaning that transient tests are generally preferred. The needle probe (or hot wire) is generally recommended for cohesive soils. Granular materials are more challenging due to the need to reconstitute to field conditions. The transient hot disk (transient plane source or its modification) is also appropriate for rocks with smooth surfaces. Despite the limitations of laboratory testing, in contrast to or in addition to TRT, the thermal conductivity of the different soil layers can be estimated. Distributed thermal response testing (DTRT), with the use of fibre optic or other distributed temperature sensors within the heat exchanger, can also be applied to determine thermal conductivity heterogeneity and particular the flow of ground water in specific horizons. However, this test typically comes with increased cost and equipment needs compared to routine TRT.

2.2.2 Detailed Hydrogeological Characterization

As mentioned above, the potential development of a thermal plume and the long-term efficiency of the EGs highly depends on the hydrogeological conditions. Their critical role imposes a deeply detailed characterization whose cost will be refunded during the project by an increase of thermal efficiency. On the contrary, a poor hydrogeological characterization is highly susceptible to induce multiyear thermal shift and, in the end, non-sustainability of the thermal harnessing.

First of all, the geologic profile should be determined through specific geotechnical investigations. Mechanical tests such as SPT, CPT or pressuremeter are not sufficient since the nature of soils and rocks are important for the thermal analysis. Continuous drilling with sampling is required and its number must be adapted to the size of the project.

The effective porosity of the soil strongly affects the thermo-hydraulic properties of soils and rocks. Literature provides correlations to evaluate it through granulometric curves. However, attention should be dedicated to the in-situ stress conditions as porosity does not only depend on soil/rock nature.

The hydraulic conductivity of the soil and its anisotropy must be determined. It can be done through laboratory tests on samples or through pumping tests (e.g. Lefranc) but in case of a heterogeneous soil, a pumping test will provide more relevant information for the scale of the project. What is more, the back analysis of in-situ pumping can provide information on the anisotropy of hydraulic conductivity (mainly due to the stratification of soil layers.

The groundwater flow characteristics must also be determined. They include piezometric levels, hydraulic gradient and predominant flow direction. The seasonal variations of these characteristics should be considered, especially in the case of very shallow EGs. All these data can be extracted from available piezometric maps but in case of a lack of information, it is crucial to install new piezometric wells and register their level for at least one year.

In the case of energy tunnels or energy diaphragm walls, it is crucial to estimate the perturbation they will induce in the groundwater flow characteristics since their large longitudinal extent will generate a dam effect and then affect seepage velocity as well as seepage direction. This drastic change of groundwater flow conditions will impact on one hand the efficiency of heat exchange and on the other hand the properties of induced thermal plumes.

The detailed hydrogeological characterization hereby presented will generally lead to quite homogeneous model. However, the subsurface often present a significant heterogeneity which can affect the thermal design of the project. It can be helpful to add geophysical methods and/or distributed temperature sensing to increase the spatial resolution (Hermans et al. 2014).

2.2.3 Detailed Thermomechanical Characterization

When designing EGs, it is important to ensure that geotechnical and structural limit states are not exceeded due to the additional stresses and strains resulting from temperature changes due to their use as heat exchangers. However, this analysis requires some additional ground geotechnical parameters and an understanding of how they are affected by the temperature changes, apart from the parameters used for typical ground-structure design.

The most important soil parameters required for geotechnical design involve soil deformability, strength, in situ stress state, stress history, overconsolidation ratio (OCR), and water regime. There are numerous reference texts on this subject (e.g., Look 2007), which refer in particular to the Eurocode 7—Geotechnical Design standard. The additional information required for the design of SGE systems, in addition to that of the geostructure itself (coefficient of thermal expansion of concrete or grout), relates mainly to the volumetric thermal behaviour of the ground, based on the coefficient of thermal expansion of the ground, and whether there is a meaningful change in the geotechnical parameters with temperature, i.e., a non-isothermal behaviour of the soil.

The range of ground temperature changes due to the operation of a thermoactive geostructure is relatively modest (Knellwolf et al. 2011), mostly ± 20 °C (Loveridge et al. 2017). In addition, the temperature changes in the ground generally occur close to the EGs and with a seasonal wavelength. For these reasons, significant changes in soil geotechnical parameters are not expected below the operating temperature of the EGs (Brandl 2006; Loveridge et al. 2017; Insana 2020). Nevertheless, this aspect should be considered depending on the site-specific conditions, climatic conditions, project complexity, expected thermal demand, and properties of the thermoactive geostructure.

The mechanical effect of seasonal thermal loads in the thermoactive geostructure depends on its interaction with the surrounding ground (Cekerevac and Laloui 2004; Di Donna and Laloui 2015). A comprehensive review of the main characteristics of soil thermomechanical behaviour associated with SGE operation and the main trends observed in the evolution of geotechnical parameters is described in Laloui et al. (2014).

Laloui et al. (2014) also addressed the issue of evaluating temperature variations at the ground-pile interface and indicated that the stresses caused by temperature variations may be negligible; however, this aspect should be further validated, and the maximum allowable temperature variations should be considered in the standards.

The volumetric expansion of the soil is of particular importance because of its interaction with the concrete structure (Loveridge et al. 2017). The relative expansion between the geostructure concrete and the surrounding soil, as well as the constraints of the geostructure, are critical factors in the stresses and strains induced in both the concrete and the soil. They also affect the deformation of energy pile groups and the thermally induced deformation of the soil (Rotta Loria and Laloui 2017).

The thermomechanical volume behaviour of soil is complex due to its highly nonlinear and irreversible stress–strain behaviour and its multiphase nature. It is not possible to define a simple soil thermal volumetric expansion coefficient (Mitchell 1993) because it depends on several aspects related to soil properties, in situ conditions (effective stresses, pore water pressure, permeability, drainage conditions), and imposed thermal loads. However, due to the difficulty of evaluating this parameter and for simplicity, a constant value is generally assumed. Typical values for solid grains range from 1×10^{-5} to 3.4×10^{-5} °C^{-1} (Delage 2013), while the thermal expansion coefficient of water is 27×10^{-5} °C^{-1}. Due to the higher value of the thermal volume expansion of water, positive pore pressures can be generated in clayey soils during heating, resulting in a decrease in the effective soil stress. This effect is more pronounced at full saturation and decreases significantly at lower saturation levels due to the presence of air in the pores. To account for this effect, numerical modelling of the overall soil/structure system must be performed.

2.3 Thermal Loading

In the design process of a new building, the energy designer provides the energy/power demand profile. This needs to be used to determine the temperature change of the fluid (the thermal design) and of the geostructure itself (used as an input to the thermomechanical design).

Thermal loading must be considered as an action that varies over time, associated with the temperature variations to which the EG is subjected in the various seasonal operating modes (heat extraction for winter heating, heat injection for summer cooling). Given the generally high thermal inertia of the structural material in which the heat exchangers are placed, the temperature variations that occur throughout the day (differences between day and night) can be neglected. Conversely, these variations can gain importance in the energy performance analysis.

The actual temperature variations in the geostructure and the thermal loading depend on the energy demand of the supplied structures and heat receivers, which in turn depends both on the type and use of the superstructure and on the climatic conditions of the area of interest. In current applications the temperature variation for shallow geothermal energy extraction is usually limited to a range of 0–40 °C (Bourne-Webb et al. 2016b; Laloui and Di Donna 2011). In some cases, higher temperatures, up to 50 °C, can be reached, either because of a non-optimised operation or because heat is recharged in the ground from waste heat or solar panels. Excessive lowering of the ground temperature must be avoided and a threshold value of 2 °C should be considered as a minimal allowable temperature to which the geostructures and the surrounding may be subjected (Brandl 2006). Lower temperatures could generate freezing of the soil and/or the concrete mix and its additives. The heat carrier fluid can reach negative temperatures but only for a small timespan. Overall, it needs to be recognised that regulation in some countries may limit ground temperatures.

For practical design applications, it is necessary to apply a simplified thermal loading to the geostructure, with respect to the real heat flow to which it is subject, in terms of type, duration, amplitude.

2.3.1 Spatial Simplifications

The geostructure is subjected to a flow of heat transferred (or absorbed), which runs through the exchanger pipe. It is necessary to make simplification choices concerning the type of thermal loading and the location within the geostructures where to apply it.

For the thermal design in the simplest and most common case, a uniform temperature distribution is applied over the entire EG, regardless of the actual position of the heat exchanger pipes (Saggu and Chakraborty 2015; Suryatriyastuti 2013; Laloui et al. 2006). This simplification is acceptable when the loop is distributed in an almost homogeneous way, as for example in energy piles. In this case, in the absence of constraints, the geostructure will be subject to purely axial stress and strain actions. Even though the temperature can be assumed as uniformly distributed along the geostructure, in practice, temperature differences over the geostructure will occur.

The thermally induced stress and strain effects arise from the difference between the geostructure and the surrounding ground temperature, which in the first 10–15 m from the ground surface is seasonally variable and therefore closer to the working temperature of the heat transfer fluid. Assuming a uniform temperature of the shallowest portion of ground, equal to the uniform and constant value reached beyond a certain depth, means overestimating the thermally induced stress effects.

In walls and tunnels pipes could be located on one side of the geostructure (on the ground side in walls and at the intrados or extrados in tunnel linings). In these cases, the temperature variation induces bending moments which would be neglected by choosing the simplification of uniform temperature in the geostructure. An accurate analysis would require applying the thermal loading in its real position.

In the more complex case, the heat exchanger pipe can be modelled in its real total development. Thermal loading can be introduced by applying a temperature distribution to the pipe itself, gradually varying between inlet and outlet. Alternatively, the fluid flow inside the pipe can be modelled, assigning the flow rate and temperature to the inlet and thus allowing the natural development of both the heat transfer process and the thermal stresses and strains in the geostructure.

2.3.2 Temporal Simplifications

Another simplification concerns the time history of thermal loading, to be applied on an annual basis over the operation cycle of heating/cooling. One can use a load that reproduces real time variations, but it can be modelled by a stepwise load or a variable sinusoidal load (Fig. 2.3). In the simplest case, a two-step thermal history can be applied, corresponding to the heating period (with a constant minimum temperature, in the winter months) and the cooling period (with a constant maximum temperature, in the summer months). More often the seasonal variability is reproduced with a sinusoidal curve, similar to the one used to represent the seasonal variability of atmospheric temperature. The hypothesis underlying this assumption is that the heat transfer fluid works at temperatures that are always proportional to the atmospheric ones, over the entire annual cycle. The amplitude of the sinusoidal curve is generally determined by the maximum summer and minimum winter temperatures expected in the area of interest.

Different structural responses can be obtained for thermal loadings that differ in time history. Both the simplified approaches described tend to overestimate the thermal loading on the geostructure, as they apply maximum and minimum temperature values over wider time intervals than those in which they actually occur (trying to get a load whose average is a correct estimate would lead to consider lower extreme values and could lead to an underestimation of potential thermal ratchetting effects). Additionally, the two-step thermal loading induces a maximum stress variation and lowest thermal strains at the beginning of the heating/cooling phase, when the temperature difference between the geostructure and the surrounding ground is maximum. This stress variation then tends to decrease during the operation phase (at constant temperature), as a consequence of the subsequent heat transfer from/to the ground. Conversely, the thermal history with sinusoidal curve induces a progressive, monotonic variation in the stress–strain state of the geostructure, as the ground is subject to a more gradual temperature variation concurrently with that of the geostructure.

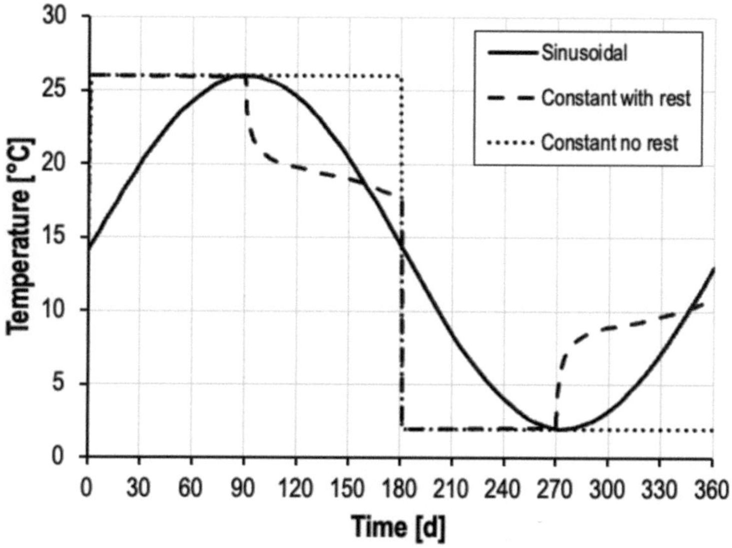

Fig. 2.3 Examples of temperature variation for different types of thermal loading functions

A more accurate analysis would require the measurement or estimation of the actual inlet and outlet temperatures of the fluid over time, and the application of an average temperature in between the two, in time steps small enough to correctly reproduce their variability. A time interval of one month is usually considered sufficient.

Several forms of thermal loading functions could be examined to assess their influence on the thermo-active geostructure reaction.

Finally, the thermal loading must be applied over several annual cycles, to analyse the transient effects of the process and both the short- and long-term behaviour. The most important of these occurs when the natural conditions of the ground fail to seasonally rebalance the system extraction or injection of thermal energy, for example when there is no significant groundwater flow, which can thermally regenerate the ground. Since the heat extraction in winter is usually greater than the heat injection in summer, a gradual depletion of natural thermal energy may be generated in the geostructures surroundings.

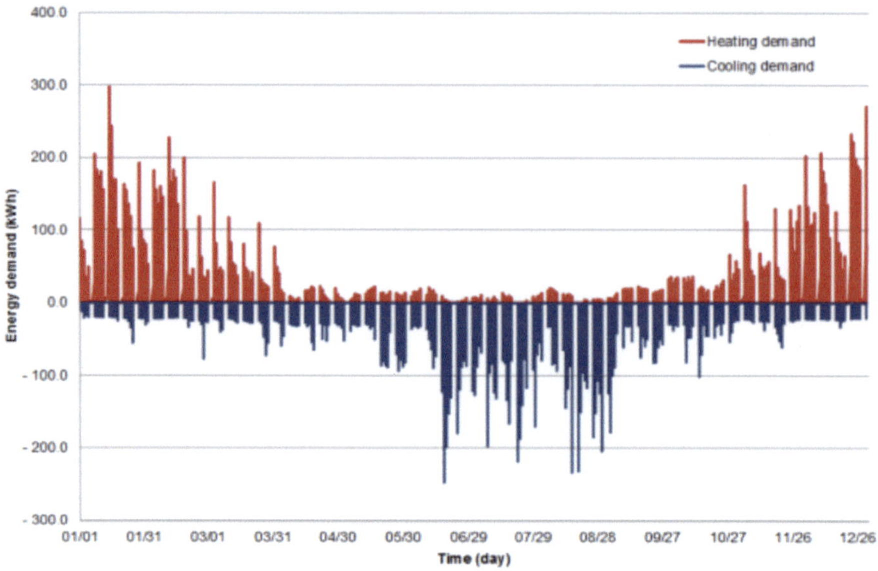

Fig. 2.4 Example of hourly energy demand in an office building

Although real temperature variations are related to daily and annual heating or cooling demands (Fig. 2.4), for design purposes, the above-described simplifications are justified.

Chapter 3
Thermal Design of Energy Geostructures

Abstract Chapter 3 presents methods for evaluating and optimizing the thermal performance of energy geostructures. It begins with feasibility assessments to estimate energy potential and proceeds to detailed design using analytical, semi-analytical, and numerical models. Different geostructure types such as piles, walls, slabs, and tunnel linings are addressed with tailored approaches. The chapter explains heat exchange mechanisms, key design parameters, and modelling techniques such as G-functions and numerical simulations. It highlights trade-offs between model simplicity and accuracy, recommending numerical methods for complex conditions. Overall, the goal is to maximize energy efficiency while maintaining acceptable temperature limits and ensuring long-term system sustainability.

Energy Geostructures (EGs) perform a dual function, one of a structural nature and the other of an energetic nature. A preliminary assessment of the energy potential, which will depend on the specific geostructure and site conditions, is essential to establish the effective efficiency of the system called upon to produce thermal energy and its sustainability, including economic sustainability.

The analysis allows one to have a rough estimate of the share of the total energy requirement that can be covered by the EG and therefore what measures can possibly be taken to supplement it.

After this preliminary assessment, a detailed energy analysis allows instead to design the geothermal exchange system in order to optimize the energy yield.

Generally speaking, the energy-optimal solution is the one that maximizes the heat exchange with the ground, while keeping the temperature changes within the circulating fluid and the ground within acceptable limits. The efficiency of the system is in fact estimated as the ability to vary the temperature of the heat transfer fluid, raising it in winter heating mode or lowering it in the alternative summer cooling mode. From an analytical point of view, the energy yield of a single exchanger tube is calculated as the heat flow Q [J/s] which is absorbed or released by the fluid along the tube, from inlet to outlet, equal to:

$$Q = \dot{m}c(T_{in} - T_{out}) \tag{3.1}$$

M. Barla et al., *Recommendations for the Conceptualisation, Design and Construction of Energy Geostructures*, SpringerBriefs in Applied Sciences and Technology,
https://doi.org/10.1007/978-3-032-30927-3_3

Table 3.1 Requirements for the different stages of the design

	Feasibility study	Outlined design	Detailed design
Foundation piles	Rules of thumb	Analytical solutions	Analytical solutions or numerical modelling
Foundation slabs	Rules of thumb	Analytical solutions or Numerical modelling	Numerical modelling
Walls	Design charts	Analytical solutions or Numerical modelling	Numerical modelling
Tunnels	Design charts or numerical modelling	Numerical modelling	Numerical modelling

where T_{in} and T_{out} are the two temperature values, $\dot{m}$ is the mass flow of the fluid [kg/s] and c is the specific heat of the fluid [J/(kg·°C)].

Equation 3.1 demonstrates that for a maximum energy yield, for given climatic and hydrogeological conditions, parameters such as the thermal properties of the heat transfer fluid, the flow rate in the pipes and the number and arrangement of the exchanger pipes can be optimized. With regard to these last aspects, it will be necessary to keep in mind the role of the thermal conditions present at the boundaries of the geostructure and the possible thermal interference between two or more exchanger tubes.

The most common methods for the preliminary assessment of the energy potential and the methods for the detailed analysis of the energy yield of the geostructure are described below. Depending on the design analysis stage, methods to be adopted may vary. Table 3.1 shows the most appropriate methods for the different stages of the design.

The analysis of the geotechnical and structural aspects, in particular of the effects induced by thermal variations on the state of stress and deformation of the EG, will instead be dealt with in Chap. 4.

3.1 Energy Piles

3.1.1 Feasibility Study

A Feasibility Study for energy piles is a strategic step which provides a preliminary assessment regarding the amount of renewable energy that the piles (that are needed in a certain project for structural reasons) can offer and the associated return of investment figures. This comprehensive study ensures that all technical, economic, and environmental factors are carefully analysed to support informed decisions.

A feasibility study provides early-stage predictability of what clients and end beneficiaries can expect by transforming standard foundation on piles or excavation retaining systems of piles into energy piles, used for heating and cooling buildings.

Conducting this study at an early stage of the project allows the project team to explore multiple scenarios for thermal activation and select the most efficient and cost-effective solution.

Key components of a feasibility study are outlined below:

a. Project Overview: Clearly define the project's objectives, scope, and energy efficiency ambitions. Identify the intended purpose of the energy piles and their integration within the overall building design.
b. Site Assessment: Perform a thorough assessment of the geotechnical and hydrogeological conditions, including soil profiles, groundwater levels, and environmental context. Identifying soil thermal properties early (using charts and correlations with geotechnical parameters) is needed to estimate energy performance. This analysis will help identify any potential challenges or site-specific advantages that may impact the overall performance of energy piles.
c. Energy Analysis: Analyse the outline energy demands of the project, including heating and cooling requirements. Assess local temperature variations as well as soil thermal properties, and future energy projections to understand the energy needs that the energy piles can cover. When the piles have the capacity to meet or exceed 100% of the building's energy demand, multiple activation scenarios can be proposed. In such cases, each scenario is supported by a cost–benefit analysis to guide optimal decision-making. If the pile system alone cannot fully cover energy needs, scenarios can be developed for hybrid systems. Options may include integrating geothermal boreholes, raft foundations, or combining energy piles with conventional systems (e.g., air-source heat pumps).
d. Thermo-mechanical aspects: Preliminary analysis of thermo-mechanical aspects need to be provided in the feasibility study to help those involved in the project have a general image and better understanding of the extent to which thermal activation of piles may or may not affect their mechanical behaviour. This early evaluation can be based on engineering experience and published data to estimate expected temperature ranges, stress changes, and deformations, ensuring safe and reliable design decisions.
e. Hydraulic sizing of the geothermal system: the full hydraulic network for circulating the heat carrier fluid and connecting the primary circuit to the heat pump needs to be designed.
f. Economic & Cost–Benefit Analysis: Perform a detailed financial analysis, considering both investment and operational and maintenance costs. Compare energy piles proposed systems with the other heating and cooling systems that are considered as alternatives for the project taking into consideration different factors and costs related to installation, exploitation and maintenance of the systems highlighting potential long-term savings and advantages. Provide a Return on Investment (ROI) analysis showing the estimated payback period for each proposed scenario.

g. Environmental Impact Assessment: Assess the environmental advantages of using energy piles, including reductions in greenhouse gas emissions and alignment with sustainability targets. This may support certification efforts (e.g., LEED, BREEAM) as well as comply with certain local/national regulations.
h. Risk Assessment: Identify potential risks and challenges associated with the implementation of EGs. Evaluate factors such as technical feasibility, construction complexities, and potential impacts on adjacent structures or infrastructure. Develop risk mitigation strategies to address identified risks.

By conducting a comprehensive Feasibility Study, project stakeholders can make informed decisions regarding the implementation of EGs, considering technical, economic, and environmental factors. This study serves as a valuable tool in assessing the feasibility and potential value of integrating EGs into the project's design and construction process. Such feasibility study is also an important input document for the design stage.

Often at feasibility stage references from other similar projects can be mentioned, especially projects from where monitoring data regarding the performance is available (as shown in Fig. 3.1). It can be seen that the heat exchange is generally in the range between 20 to 100 W/m. The differences might depend on the depth of the installation, the configuration of the pipes, the thermal load induced by the heating system and on soil thermal and hydraulic properties and underground conditions, especially the presence of groundwater flow.

3.1.2 Outlined and Detailed Design by Analytical Models

When analytical methods are adopted for energy piles one of two approaches are possible: (i) separation of the temperature response within the piles and the ground; (ii) calculation of a combined temperature response. In the former approach, the temperature change of the fluid circulating in heat transfer pipes of the pile, ΔT_f is expressed as:

$$\Delta T_f = qR_b + \frac{q}{2\pi \lambda_g} G(r, t) \tag{3.2}$$

where q is the applied heat transfer rate in Watts per metre depth of the pile, R_b is the pile thermal resistance (mK/W), λ_g is the ground thermal conductivity (W/mK), and G is a function of the elapsed time t (s) and the radial coordinate r (m). G-functions are generally used to describe the temperature change in the ground in the vicinity of a ground heat exchanger (GHE) due to the applied heat transfer rate q. In Eq. 3.2, the two terms of the right-hand side can be evaluated separately thus simplifying the analysis. However, it is based on a steady state being present within the pile and only accounts for transient conditions in the surrounding ground. For approach (ii) then the heat diffusion equation is solved transiently for the pile and ground together.

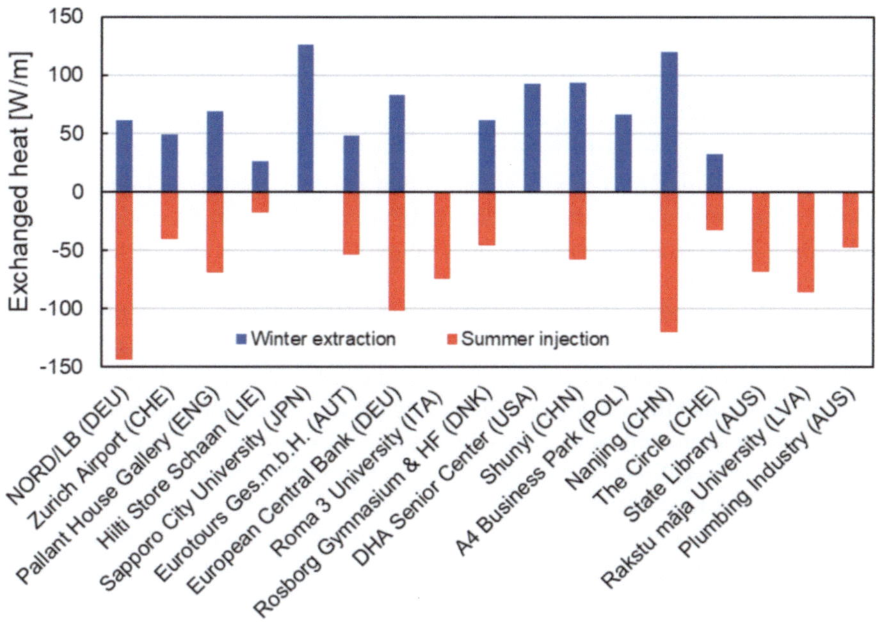

Fig. 3.1 Heat exchange per meter of pile length in real energy piles (from Cost Action Foliage, Salciarini et al. 2026)

For consideration of a single pile, the radial coordinate in Eq. 3.2 is equal to the pile radius r_b. However, piles occur in pile groups and to account for thermal interactions between piles it is also necessary to determine the temperature change at other values of r, equal to the spacing between the piles. Then the overall temperature change can be determined by spatial superposition.

Furthermore, in operation, q usually varies in time, and therefore temporal as well as spatial superposition will also need to be applied to Eq. 3.2 (e.g. Spitler and Bernier 2016). Because application of superposition for any number of piles and for decades of small timesteps of thermal demand soon make the analysis time consuming from first principles, it can be preferable to use software tools for this task. However, the majority of these software tools are developed for use with BHEs which have some different characteristics to energy piles, despite their superficial similarity (e.g. typically close to axisymmetric).

G-functions play a fundamental role for analytical models on the thermal design of energy geostructures such as energy piles. A detailed review on these G-functions and their application to energy piles can be found in Loveridge et al. (2020b). Some possible analytical models for application are summarised below and in Table 3.2 and reviewed with respect to their suitability. It should be noted that all models will assume that the ground behaves as a single homogeneous and isotropic medium. While this assumption is not always a suitable reflection of reality, it can to some extent be dealt with through the use of averaged or lumped thermal parameters. For significantly layered or heterogeneous soils, however, this assumption must at least be acknowledged since heat transfer will not be constant with depth as is explicitly implied by this assumption. Where ground water flow is present, temperature change will be overestimated, sometimes significantly, and recourse to more computationally expensive numerical modelling may be required.

The simplest analytical model assumes that the pile acts as an infinitely thin, infinitely long heat source, the so-called infinite line source (ILS) model (Carslaw and Jaeger 1959). Since it neglects the pile geometry (diameter and length) it is recommended to use more advanced models. The infinite cylindrical source model represents the pile as heat source which is a hollow cylinder with a finite radius but still neglecting its finite length, ground surface boundary conditions or the heat capacity of the piles material (e.g. Carslaw and Jaeger 1959; Ingersoll et al. 1954). It has been shown that this model, although being present in certain software packages, will overestimate the temperature response for energy piles (Loveridge and Powrie 2013b). The solid cylinder model would in fact be more appropriate, further details of which are found in Man et al. (2010).

To incorporate the ground surface boundary condition, the finite line source (FLS) model can be used. It assumes the upper end of the GHE is bounded by a constant temperature surface equal to the average annual air. Furthermore, it assumes undisturbed ground temperature, and the lower end of the pile to be embedded in the ground. There are various analytical solutions for the FLS of which the version by Lamarche and Beauchamp (2007) is recognised to be the fastest computationally.

While the above-mentioned models are purely analytical and determined for the solution of the diffusion equations for different geometric conditions, there are also empirical analytically implemented G-functions derived especially for piles, e.g. from Loveridge and Powrie (2013a). The pile G-functions have been developed using numerical simulation to the same surface boundary condition as the FLS model. There are many other models available in the literature, see Fadejev et al. (2017) or Loveridge et al. (2020b). When choosing a model, the following key criteria should be assessed.

- ease of use. Analytical models are chosen since they have faster run times than numerical approaches. However, ease of use still varies between use cases. This is relevant if preparing outline design calculations from first principles but is less relevant if commercial software is used.
- pile specific geometry. Models based around line and cylindrical sources do not capture the detail of the pile itself. This distinction means that there can be discrepancies in the short-term temperature response, especially if short time step analysis is required.
- appropriate pile head boundary conditions. Most models are either "infinite", assuming there is no pile head or toe and therefore temperature changes will happen ad infinitum under constant thermal demand, or "finite" which assumes a constant temperature ground surface and finite pile length meaning temperature change will reach a steady state in the long term under constant thermal demand. Most piles are actually more akin to semi-infinite conditions, since

Table 3.2 Suitability of different analytical models for the thermal design of energy piles

Model	Ease of use	Pile specific geometry	Pile open to the air	Pile beneath a building	Multiple piles	Transient pile behaviour
Infinite line source	✓✓✓	✗	✗✗	✓✓	✓✓✓	NA
Infinite cylindrical source	✓	✗✗	✗✗	✓✓	✓✓	NA
Finite line source	✓	✗	✓✓✓	✗✗	✓✓	NA
Pile G-functions	✓✓	✓✓	✓✓✓	✗✗	✗	NA
Semi-finite cylindrical source	✓✓	✗✗	✗	✓✓✓	✗	NA
Transient radial model	✓	✓✓	✗✗	✓✓	✓✓	✓✓✓

they can be considered to have an insulated pile head where they are located beneath a building. In this case an infinite model will overestimate the temperature change and the finite model will underestimate it. However, it is expected that the discrepancy will be lesser for the infinite case, especially under the scenario of varying thermal demand. Maragna and Loveridge (2019) provide an empirical approach for a semi-infinite solution specifically for piles, derived from numerical simulation.

- applicability to multiple piles. To be applicable to pile groups an analytical model must be suitable for evaluation at different radial coordinates. This is possible with purely analytical models. However, it is not straightforward with approaches derived from the outputs of numerical simulation.
- transient pile behaviour. Only the transient radial model (Loveridge et al. 2022) is applicable to address this aspect.

As seen in Table 3.2, there is no single analytical model which meets all of the key criteria perfectly. The final choice of approach will therefore depend upon factors including: the design stage, the diameter of the pile and number of pipes installed, the aspect ratio of pile, the nature of the thermal demand, the length of the required timesteps of analysis, and the design life required.

Beside the G-function, the thermal resistance plays a crucial role in Eq. 3.2, since it implies a steady state condition within the pile. Although not appropriate at short time steps (e.g. analysis with time steps of less than one day), the concept is still useful.

By application of the multipole method Claesson and Javed (2020) have developed an accurate expression for the thermal resistance of energy piles. In Eq. 3.3 the total resistance is given by two terms, the first representing the resistance of the pipes and fluid circulating within them, and the second giving the resistance of the concrete part of the pile (R_c).

$$R_b = \frac{R_p}{N} + \frac{1}{2\pi N\lambda_c}\left[ln\left(\frac{r_b^N}{Nr_p r\prime^{(N-1)}}\right) + \sigma\, ln\left(\frac{r_b^{2N}}{r_b^{2N} - r\prime^{2N}}\right)\right] \tag{3.3}$$

where N is the number of pipes installed at equal spacing around a virtual circle of radius $r\prime$, R_p is the thermal resistance of the plastic and fluid in a single pipe, and r_p is the outer pipe radius. σ accounts for the ground and concrete conductivity ratios and is given by $(\lambda_c - \lambda_g)/(\lambda_c + \lambda_g)$.

Claesson and Javed (2020) show Eq. 3.3 to be valid for all energy piles with a diameter of 600 mm or more providing the pipe diameter is no more than 40 mm. It is also applicable to smaller energy piles, within certain pipe size and number ranges.

For analysis with time steps of less than one day using the steady state resistance will overestimate the temperature change. In such cases a transient approach to resistance can be taken, replacing the term qR_b in Eq. 3.2, with $qR_bG_c(t)$ where G_c is a transient G-function giving the proportion of the steady state resistance that has been achieved as a function of time.

$$G_c = a\left[ln\left(\frac{\alpha_g t}{r_b^2}\right)\right]^6 + b\left[ln\left(\frac{\alpha_g t}{r_b^2}\right)\right]^5 + c\left[ln\left(\frac{\alpha_g t}{r_b^2}\right)\right]^4 + d\left[ln\left(\frac{\alpha_g t}{r_b^2}\right)\right]^3 + e\left[ln\left(\frac{\alpha_g t}{r_b^2}\right)\right]^2 + f\left[ln\left(\frac{\alpha_g t}{r_b^2}\right)\right] + g \tag{3.4}$$

Equation 3.4 is based on curve fitting of the output of numerical simulation and the curve fit constants a to g can be found in Loveridge and Powrie (2013a).

Alternatively, it is possible to solve the heat diffusion equations transiently for the pile and ground domains together in a combined thermal response analyses.

A radial model where both the GHE material and the ground are solved transiently assuming no boundary effects was proposed for borehole GHE (Claesson and Javed 2011; Claesson 2011) and tested for energy piles (Loveridge et al. 2022). This truly analytical approach simplifies the GHE geometry to an annulus, yet does so in a way linked to the original GHE geometry. It is very accurate for small diameter GHE like boreholes, and sufficiently accurate for piles. Like the ILS, end effects of the GHE are neglected, but the pile thermal capacity is accounted for fully. It has yet to be tested for piles larger than 1800 mm.

There are also software tools available that use pre-programmed configurations for multiple borehole GHE (Liu et al. 2021). These typically rely on underlying models like the infinite or finite line source, or the infinite cylindrical source. They can be used for energy piles with an understanding of their limitations. An additional factor to consider is the layout of multiple piles. Some software programmes use pre-selected regular arrangements for typical boreholes fields, whereas piled foundations may have more irregular arrangements depending on the design of the structure and positioning of the main structural components.

3.1.3 Outline and Detailed Design by Numerical Methods

In the detailed study, numerical methods can be advantageous since they can overcome the limitations of many analytical or semi analytical models. Numerical models can capture accurately aspects like the pipe dimensions and their location in the pile, the heat transfer between pipe, pile and soil and soil layering (see reviews of Mohamad et al. 2021 and Aresti et al. 2018). However, analytical models are faster, and numerical models require in depth expert knowledge to allow accurate predictions.

A numerical model for the analysis of the thermal performance of single energy piles or energy pile groups should clearly avoid a long-term drift of the undisturbed ground temperature by keeping the cumulative warming or cooling within a few kelvins for decades, e.g. 20 years. To ensure sustainability and long-term capacity by assessing both short-term deliverable power and long-term ground temperature drift, the simulation has to cover multiple design years including production seasons and off-periods for thermal recovery. Evidence from monitored systems and calibrated simulations shows that, besides layered ground, seasonal balance dominates the ultimate temperature envelope around the piles (Bernier et al. 2008; Zhang et al. 2021).

The precise representation of the soil stratigraphy has a major impact on the energy assessment. Thermal soil parameters, possibly also reflecting anisotropy, and groundwater conditions are first order input parameters for the individual layers. In-situ TRT can deliver representative thermal conditions. Performing the detailed design purely based on material properties gained from the literature, without any in-situ or laboratory testing, is not recommended. Careful modelling of the hydraulic parameters such as ground water level, flow direction and velocity as well as groundwater temperature is having a significant impact on thermal simulations (Li et al. 2022; Magraner et al. 2021).

Besides the soil properties, the consideration of realistic model boundaries is essential for a good prediction. The numerical mesh should extend several diffusion lengths for the simulation horizon so that the far-field boundary condition does not contaminate results. In case of a single pile an horizontal extension around 15–20 times the diameter of the pile is appropriate, while for a group of piles an extension of 2 to 3 times the width of the pile group can be adopted. For the vertical direction typically at least one time the length of the pile can be used.

As thermal boundary conditions at the upper edge it is recommended to have the most recent external temperature variation data being imposed as a Dirichlet or Cauchy boundary condition on the surface, to obtain a temperature profile that varies with depth. For the lower domain boundary, the temperature is usually fixed at the undisturbed ground value and the geothermal heat flux is neglected. However, if the domain depth shall be limited to safe computational costs, a geothermal heat flow can be imposed (Neumann type boundary condition). The lateral boundaries of the domain are usually considered adiabatic.

The real surface or basement conditions and the geometry of slabs or underground structures that can act as heat sinks or sources should be modelled, although their impact on the overall thermal response of the energy pile system is less pronounced than the impact of soil and hydraulic conditions. Cecinato and Loveridge (2015) demonstrated that pile spacing, the circuit configuration and the concrete conductivity have much stronger effects on thermal efficiency than, for example, a modest increases in fluid velocity. Higher concrete-conductivity generally lowers the field thermal resistance, while wider pipe spacing within the cage reduces thermal short-circuiting and improves fluid temperature under load. As for other energy geostructures, it is possible to neglect the thermal diffusion in the thickness of the exchanger tubes to save computational time and to simplify the analysis.

If possible, the 3D numerical simulations should be calibrated based on in-situ TRT, preferably on a test pile. Parameters such as the thermal conductivity, initial ground temperature, ground water conditions and exchanger pipes cross sectional details may be adjusted if they are based on assumptions rather than measurements. A robust numerical model also needs to be checked for being mesh and time-step independent. Furthermore, such models can be used to verify computationally faster models like empirical or semi-empirical models which may be used for example for parametric studies (Jensen-Page et al. 2019).

The numerical simulations allow an iteration to try to meet the target energy goals by minimizing construction complexity and costs. As an example, within a pile group the analysis could highlight the number and location of the piles to be activated and the pipe circuit geometry within them.

Finally, the simulations can be used to control the limits for minimum and maximum fluid temperature to prevent freezing or overheating while still delivering the building energy load. A near-balanced annual energy budget is preferable, since a cumulative imbalance is the primary driver of multi-year thermal drift and an eventual loss of thermal capacity.

The recommendations above are given for the thermal design of energy piles and pile groups. Many of them are also applicable for thermo-mechanical coupled design (see Sect. 4.1). Especially in the case of energy pile groups, the thermo-mechanical interaction between active and passive piles has to be considered carefully, as shown by Di Donna et al. (2016) or Mohamad et al. (2021).

3.2 Energy Walls and Slabs

Most design methods for planar EGs have been developed with energy walls in mind. Energy slabs, often installed in the same solution as walls have received little specific study. However, their geometry and boundary conditions are very similar to the exposed part of an embedded energy wall and their heat transfer characteristics will therefore be largely analogous. Therefore, the below deceptions, which are developed for energy walls, can be taken to be in the most part also applicable to the thermal analysis of energy slabs. However, care should be taken with an anisotropy in the ground conditions that could impact the heat flow and any details where the slabs and walls are connected.

3.2.1 Feasibility Study

A preliminary evaluation of the heat exploitable by energy walls can be given by looking at previous existing installations. Figure 3.2 shows the heat exchanged in winter and summer in terms of W/m^2 for a number of existing energy walls installations (Di Donna et al. 2017a; Salciarini et al. 2026). It can be seen that the heat exchange is generally in the range between 10 to 50 W/m^2, with some exceptions. It is anticipated that the differences might depend on the depth of the installation, the configuration of the pipes, the thermal load induced by the heating system and, as for the case of energy piles, on soil thermal and hydraulic properties and underground conditions, especially the presence of groundwater flow.

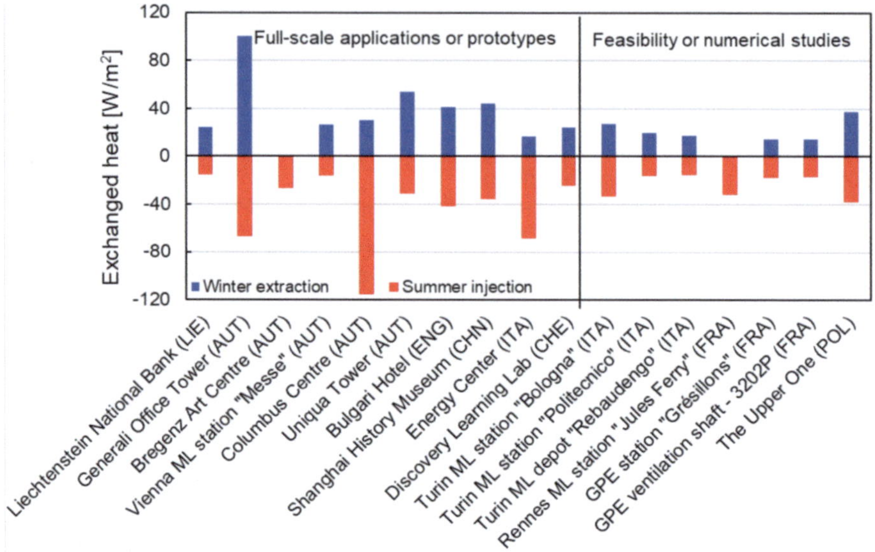

Fig. 3.2 Heat exchange per square meter of wall surface (from Cost Action Foliage, Salciarini et al. 2026)

An additional aspect which must be considered specifically for energy walls (and tunnels), that is not necessary with energy piles, is the effect of temperature and speed of the air inside the underground space (Nicholson et al. 2013; Nicholson et al. 2014; Bourne Webb et al. 2016a; Di Donna et al. 2017b; Makasis et al. 2020; Sterpi et al. 2020). These energy geostructures will exchange heat not only with the ground, but also with the air inside the underground space. This is discussed further in Bourne Webb et al. (2016a) and in Makasis et al. (2020) where they show that assuming insulated boundary conditions is a conservative choice for design. In the case of e.g. metro station excavations which contain hot air from train breaking this might represent an additional source of heat during winter, but a drawback during summer. However, other excavation where there is cold, stationary air, this may represent a reduction in heat extraction capacity.

For the analysis of the geothermal potential in this type of EGs, it is possible to resort to literature nomograms which, summarizing the results of numerical parametric analyses, allow to estimate the thermal power per unit of surface, in the winter heating and summer cooling modes, depending on the variation of some parameters: thermal properties of the materials, speed and direction of the flow, inlet temperature of the heat transfer fluid, duration of the process (Di Donna et al. 2021; Zannin et al. 2022). The nomograms are based on reference models and therefore their use must be limited to projects whose functional characteristics correspond to those of these models.

Due to the limited number of available data on operational energy walls, outline design charts are also presented in Fig. 3.3 (Di Donna et al. 2021). The charts are

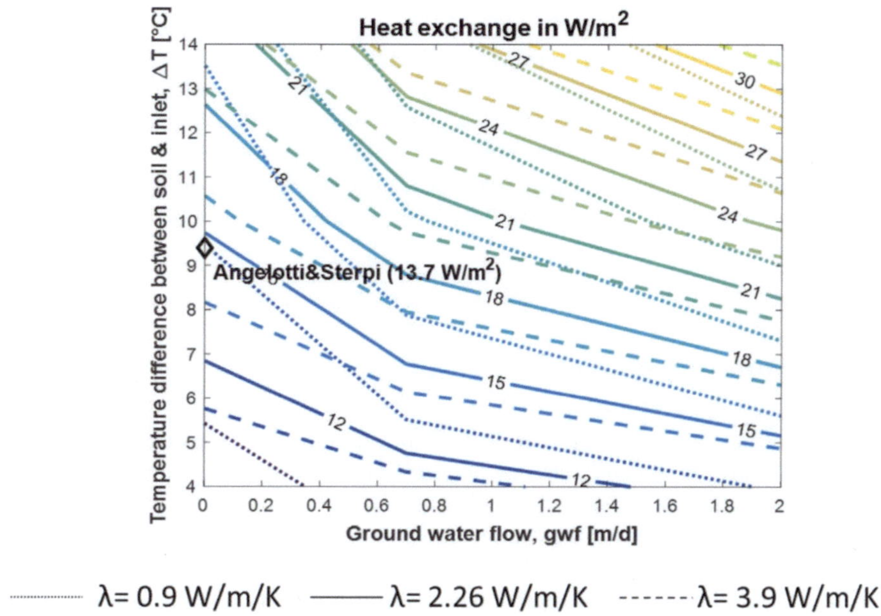

Fig. 3.3 Extracted heat in W/m^2 for energy walls (Di Donna et al. 2021)

based on a 3D model of one diaphragm wall panel, with the following specific assumptions:

- the soil is saturated and groundwater flow is assumed to be perpendicular to the wall;
- the wall panel is 20 m high, 2.5 m long and 1 m wide, with an excavation depth of 10 m on one side;
- the external air temperature is taken into account and fixed to 30 °C and 2.5 °C for summer and winter respectively;
- a constant temperature boundary condition (20 °C for summer and 10 °C for winter) is adopted for the excavation side of the wall. This assumption is more representative of applications where high airflow persists;
- pipe spacing is 50 cm, with the arrangement of the pipes following a double-U configuration, to a depth of 19.75 m. There is a lateral cover of concrete of 0.5 cm. Pipes are included on the soil side of the wall only;
- each analysis case is run for a period of 1 month.

Based on the specific site conditions, i.e. ground temperature, soil thermal conductivity and groundwater flow velocity, the charts give to the designer a reasonable preliminary idea on the heat that can be potentially extracted from the ground in winter and injected during summer, expressed in Watt per square meter of the wall. Reasonable values of heat extraction/injection range between 12 and 40 W/m^2, with maximum values reached only in the presence of significant groundwater flow and during summer. Note that special attention should be payed to underground water flow direction in comparison to the wall orientation. Ground water flow along the wall length could lead to significantly higher heat transfer. However, the complex geometry of many underground infrastructure schemes can also lead to damming of groundwater which would influence the local conditions.

The nature of the underground space will also affect the results. Where static air flow conditions are present then heat transfer rates will be lower, but only by approximately 2 W/m^2 on average.

3.2.2 Outlined and Detailed Design

For the thermal design of an energy wall, in addition to the methods described above, there are some analytical and semi-analytical techniques available together with numerical modelling.

Analytical techniques for the routine analysis of thermal performance of energy walls are still under development. The first analytical solution based on heat conduction with the addition of a convective heat transfer boundary condition at the excavation face was proposed over and under the excavation line and validated against experimental data by Sun et al. (2013). Initial semi-analytical calculation models based on thermal resistances, were developed and validated by Kürten et al. (2015) and by Shafagh and Rees (2019). A composite model based on the Dynamic Thermal Network approach was then developed and validated by Shafagh et al. (2020), giving the temperatures in the ground, the pipes and the excavation.

Probably the most useful approach due to its user-friendly nature, is the application of the infinite plane source (IPS) coupled with a resistance model, recently pioneered by Gupta (2024). This follows a similar approach to that described for piles in Paragraph 5.1.2 whereby the heat transfer processes are split into a steady and a transient component.

$$\Delta \mathrm{T}_{avg} = \frac{q_p}{N\lambda} G(x_\omega, t) + q_\omega s(R_W) + q_p s\left(R_P + R_f\right) \tag{3.5}$$

where q_p is the planar heat flux through the wall, N is the number of times the IPS is evaluated, λ is the thermal conductivity of the ground, $G(x_\omega,t)$ is the g-function for the IPS, x_w is the distance from the heat transfer pipes centre line to the wall surface where the IPS is evaluated, s is the heat transfer pipe spacing, and R_W, R_P and R_f are thermal resistances of the wall, pipes and fluid respectively.

The first term in Eq. 3.5 is transient and determines the temperature on the back of the wall, and the subsequent terms the temperature difference between the wall and the exterior of the pipes, and across the pipes respectively. Equation 3.5 needs to be applied separately to the exposed and embedded parts of the wall and then a length or depth averaged approached adopted to determine the average fluid temperature. For exposed sections of the wall, the IPS is evaluated twice, once assuming heat flow directly to the back of the wall from a single row of pipes and once assuming heat flow in the opposite direction which returns from the exposed face of the wall to the back of the wall due to an assumption of insulation. For the embedded part of the wall, heat flows into the ground to the front and back of the wall and the IPS only needs to be evaluated once on the side closest to the single row of pipes. Further details of this approach and expressions for calculation of the thermal resistance of walls with one row of pipes are contained in Gupta (2024).

The IPS approach is simple to apply but is limited to certain geometries (i.e. one row of pipes, insulated exposed face of the wall) and timescales (correction factors are required in the long term). Considering these simplifications on the timescale of analysis and the geometry under consideration, a numerical analysis of the specific case of interest is recommended. This allows to simulate the heat exchange process on the real geometry of the work and with the real working conditions.

In fact, despite the greater complexity of setting and the lengthening of calculation times, only numerical modelling allows to grasp fully the complexity of the geometry of the structure and the surrounding conditions, to the point of allowing, in the most refined models, to take into account the real arrangement and method of connection of the heat exchanger tubes, of the two- and three-dimensional effects and of any couplings between thermo-hydro-mechanical physical processes.

The numerical analysis of the energy support structures requires at least a two-dimensional geometry, in which a cross section of the geostructure and the surrounding terrain is analysed, assuming flat heat diffusion conditions. If the arrangement of the pipes along the structure is irregular and/or if the latter is not particularly extended longitudinally with respect to the dimensions of its cross section, it will instead be advisable to resort to a three-dimensional model of the work.

The geometry and extent of the terrain domain to be included in the simulation depend on the problem considered. The typical domain shape for a 2D or 3D analysis of an energy support structure will be rectangular or prismatic, respectively. The extension (lateral and in depth) of the domain will depend on the overall simulation time and on the thermal properties of the terrain, therefore it is recommended to resort to a preliminary numerical experimentation to establish the best compromise between domain dimensions and computational times. For a purely diffusive simulation lasting on the order of a week, a horizontal extension equal to 15–20 times the thickness of the energy support structure and in depth equal to 1/3 of its length could be considered sufficient. Alternatively, the domain size can be reduced by adopting adiabatic lateral boundary conditions in case of symmetry in the heat exchange, or by calibrating open boundary conditions.

Typically, in the case of an extended domain, the thermal boundary conditions to be imposed at the external boundaries of the terrain are of the Dirichlet type, i.e. a constant temperature equal to the undisturbed value. For the lower domain boundary, in certain cases (e.g. for very deep installations or in areas with anomalous geothermal gradients) a geothermal heat flux may be imposed (Neumann type boundary condition).

The choice of the boundary conditions on the side of the excavation is more complex and depends both on the conditions of use of this space and on the type of fluid (air or water) present at the interface with the structure. In the case of bulkheads, diaphragms or sheet piles which separate the ground from a closed underground space, the boundary condition which represents the interface with the air may concern the imposition of a temperature (constant or variable over time) or a convective heat flow. The first case is typical of underground car parks (often at a temperature similar to atmospheric temperature) or of underground stations (often at a temperature dictated by the air conditioning system of the station itself), the second case may concern railway tunnels where there is a significant presence of air circulation.

In the latter case the choice of the convective heat exchange coefficient is not trivial. On the side of the excavation, a convective flow is often considered by imposing a Cauchy boundary condition in which it is necessary to define the temperature time history (constant or variable) of the internal air and a conductance parameter h. In this regard it is possible to find guidance in the ISO 6946 (2017) standard, where conductance values are suggested for different types of internal and external spaces. Typical values for h vary between 6 and 20 $W/(m^2K)$, although values of h in the order of 50 $W/(m^2K)$ for air stream speeds of 10 m/s are suggested. A literature review of examples of conductance selection in tunnels of various types can be found in Di Donna et al. (2017b).

The boundary conditions at the upper boundary of the domain will also depend on the case considered and may include a time series of temperatures similar to atmospheric ones in the case of outdoor exposure, an adiabatic condition in the presence of thermal insulation or a temperature equal to that of the space inside the building bordering the simulated domain.

To accurately reproduce the heat transfer in energy walls and slabs, the initial conditions also deserve attention. If the simplest choice is to assume, at thermal equilibrium, an initial temperature that is constant in the domain and equal to the undisturbed one of the ground, the accuracy of the simulations improves considerably by imposing an initial temperature that varies with depth and is equal to that measured in several points of the domain before the geothermal activation of the work. In the absence of field measurements, it is advisable to resort to preliminary simulations in which, for some annual cycles, a thermal history can be imposed on the upper (air) and lateral (air or water) boundaries of the domain as well as an initial temperature distribution in the ground. A simplified way, although not so effective for energy walls and slabs, is represented by the imposition of Hillel's analytical function (Hillel 2013) which describes a temperature profile in an infinite half-space of the ground as a function of the atmospheric one.

Some authors have recently studied the heat transfer in thermoactive diaphragm walls through detailed numerical models and have identified the most important design parameters in maximizing their energy efficiency. Sterpi et al. (2020) identified the arrangement of the heat exchanger pipes and the length of the section embedded in the ground as determining factors, concluding that the energy performance can improve by limiting the thermal interference between the sections of pipe carrying the fluid at different temperatures and by maximizing the heat exchange in the deepest section of the structure, in contact with the ground on both sides. Di Donna et al. (2017b) through parametric analyses have compiled a ranking of the most influential parameters in both short- and long-term functioning. The results suggest that increasing the number of pipes for each panel is the best strategy to maximize the energy efficiency of the diaphragm walls in the short term, while in the long-term operation the temperature difference between the air and the ground becomes dominant. This confirms the considerable benefit that is obtained by exploiting the waste heat present in the underground tunnels (mainly coming from the train braking systems) in the winter operation.

3.3 Energy Tunnels

3.3.1 Feasibility Study

A preliminary evaluation useful at the pre-feasibility or feasibility stage of the heat exploitable in winter heating and summer cooling modes thanks to the thermal activation of a tunnel lining can be given by looking at previous existing cases which have documented thermal exchanges generally between 10 and 75 W/m^2 (Di Donna et al. 2017a; Insana 2020; Salciarini et al. 2026). The wide range of values shown in Fig. 3.4 is mainly dictated by the specific thermo-hydrogeological conditions of the site which, under favourable conditions (e.g. moving groundwater), can significantly improve the heat exchange. Furthermore, like energy support structures, tunnels are also characterized by some peculiarities linked to their particular geometry which makes it important to evaluate the direction of the groundwater flow and the thermal condition of the intrados (in particular, the speed and temperature of the air flow).

Given the limited data available on operational energy tunnels, it is very useful to refer to nomograms proposed in the literature on the basis of numerical simulation analyses. An example is reported in Insana and Barla (2020), who after validating a numerical model on the basis of tests carried out on a prototype on site, have provided an overall estimate of the thermal performance in a number of different conditions in terms of direction and groundwater flow velocity, soil temperature (and corresponding internal air temperature), total thermal conductivity of the soil. The nomograms thus obtained, illustrated in Fig. 3.5, show the thermal power in W/m^2 for the winter and for the summer case for different combinations of the parameters indicated above.

Based on the specific site conditions, i.e. ground temperature, soil thermal conductivity and groundwater flow velocity, the charts give to the designer a reasonable preliminary idea on the heat that can be potentially exchanged from the ground in winter and injected during summer, expressed in Watt per square meter of tunnel lining. Reasonable values of heat extraction/injection range between 10 and 80 W/m^2, with maximum values reached only in the presence of significant groundwater

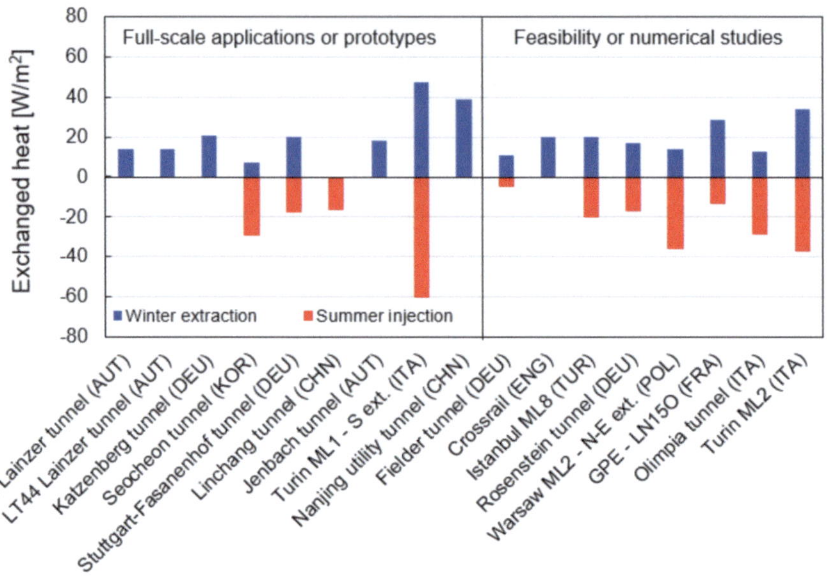

Fig. 3.4 Heat exchange per square meter of tunnel surface (from Cost Action Foliage, Salciarini et al. 2026)

flow. Note that special attention should be payed to underground water flow direction in comparison to tunnel orientation, which can significantly affect the results.

The use of nomograms must be limited to projects whose characteristics are similar to those of the models from which they originated. The nomograms, in fact, refer to the "Enertun" configuration of the geothermal probes, having a main circumferential direction, perpendicular to the tunnel axis. The depth of the tunnel in the calculation was assumed such that the influence of external air temperature fluctuations could be neglected. The results refer to a thermal activation period of 30 days. This assumption is conservative because it is assumed that the thermal power exchanged is higher in the previous days. Nonetheless, it should be noted that if the thermal activation lasts more than 30 days, especially in the absence of groundwater flow, the thermal power will tend to decrease or, in any case, to reach an equilibrium situation more slowly and later. In this circumstance the values obtained from the nomograms can reasonably be assumed as mean values of powers.

As it is possible to observe, regardless of the direction of the groundwater flow, the best performance is obtained in the case of maximum ground thermal conductivity, maximum groundwater flow speed, thanks to the thermal recharge mechanism which allows the ground to return more rapidly to the initial undisturbed, and maximum ground temperature in winter and minimum in summer. As groundwater flow velocity decreases, thermal conductivity starts to play a role. The effect of the groundwater flow direction is to significantly improve the energy performance in the case of oblique and perpendicular flow.

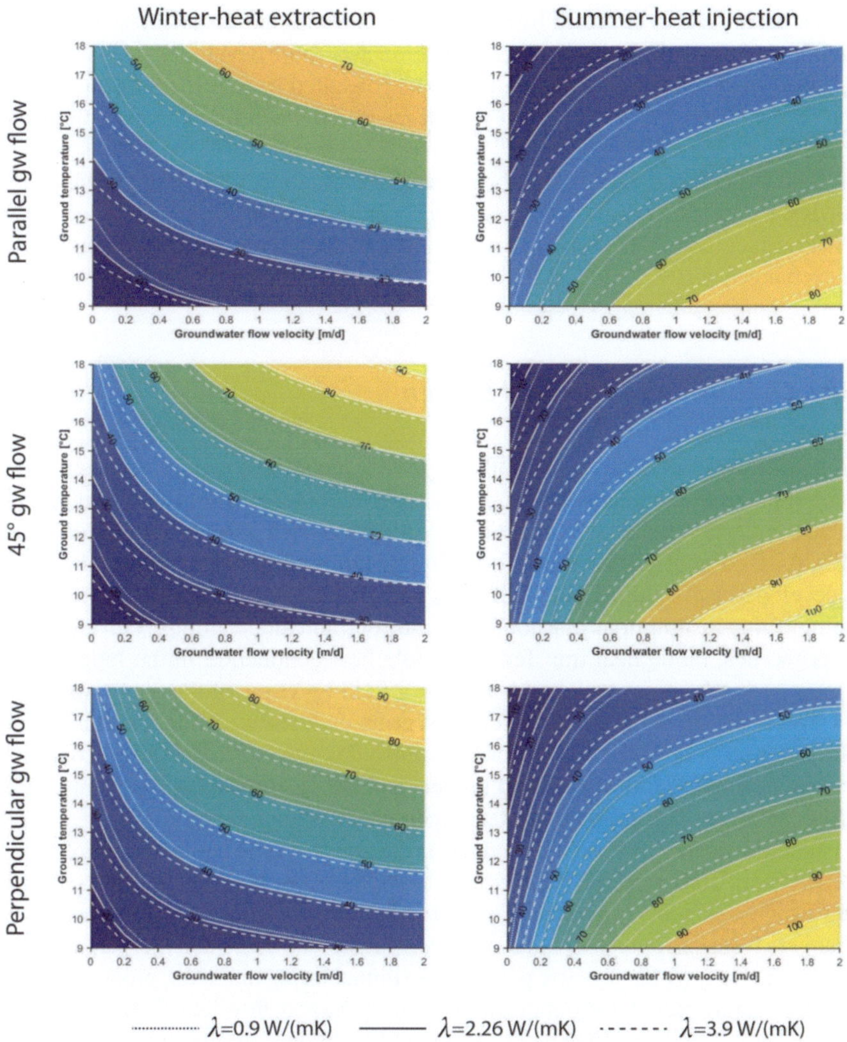

Fig. 3.5 Design charts for the evaluation of the geothermal potential in W/m^2 in winter and in summer for different directions of groundwater flow with respect to the tunnel axis (0°, 45° and 90°) (Insana and Barla 2020; Insana 2020)

Among the most influential parameters on the quantity of exchangeable heat it is therefore possible to mention the undisturbed temperature of the ground relative to the fluid temperature, the hydraulic gradient combined with the hydraulic conductivity of the ground (Darcy velocity), the direction of the groundwater flow, the air temperature in the tunnel. Furthermore, the thermal properties of the concrete lining and of any filling on the back must be taken into consideration, which in the aforementioned nomograms are set at 1.5 and 2 W/(mK) respectively.

Other design charts obtained from numerical analyses that focus more on the role of indoor air are also available in the literature, such as those developed by Dornberger et al. (2022) and Ma et al. (2021) which can be consulted if the role of the internal air of the tunnel is considered to prevail with respect to the hydrogeological conditions of the site.

It should be noted that more sophisticated evaluation tools, described in the following Chapters (in particular numerical analyses), and a more detailed study are instead recommended in the subsequent design phases to accurately simulate the heat exchange process.

3.3.2 *Outlined and Detailed Design*

The classical line heat source and hollow cylinder model adopted for BHE and piles are not applicable to tunnels. Research on analytical methods for energy tunnels is currently underdeveloped. An analytical solution has been proposed by Zhang et al. (2013). A 2D composite medium multiple line heat source model in radial coordinates was developed and validated on TRT tests performed in Linchang tunnel, based on the simplifying assumptions of a homogeneous rock and tunnel lining, with thermally inert properties, radial heat transfer, negligible heat conduction in the longitudinal direction nor in the pipes wall and heat carrier fluid by virtue of the pipes small size compared to the tunnel domain, absence of groundwater flow. The model accounts for the internal boundary condition via a sinusoidal varying temperature condition determined from monitoring of road tunnels and allows to compute the heat carrier fluid temperature. The approach has not been adopted more widely.

Less sophisticated methods, though accurate and less time demanding, are represented by semi-analytical methods which derive the thermal resistance of the tunnel lining heat exchangers. Investigations into thermal resistance of the tunnel heat exchangers were carried out by Shafagh et al. (2020) who evaluated the shape factors with varying ranges of geometric parameters (tunnel lining diameter, pipes diameter, concrete cover) to predict the outlet temperature or the heat flow rate. This approach is even more simplified than the analytical method mentioned above, but much simpler to adopt in routine application. It has to be reminded that thermal resistance is a steady-state parameter and would hence lead to underestimation of energy availability. Transient methods are recommended for tunnel heat exchangers design if possible. Therefore, although, as shown, some examples of analytical methods exist in the literature, able to capture the thermal performance of tunnel heat exchangers, most of the assumptions bring limitations. In response to this, it is more common for the tunnel and surrounding ground to be treated by numerical simulation. Numerical models solving the governing heat transfer equations are the most common tool to deal with the analysis of thermal performance. In general, it is possible to refer to finite element or finite difference calculation codes, constructed by adopting a TH (Thermo-Hydraulic) or THM (Thermo-Hydro-Mechanical) formulation.

Typically modelling is conducted in three dimensions with 1D pipes to reduce computational effort. It is advisable to resort to a preliminary numerical trial to establish the best compromise between domain extension and computational times. Indicatively, a horizontal extension equal to 15–20 times the diameter of the energy tunnel and a vertical extension equal to 8–10 times the same diameter could be considered sufficient. In some cases, such as for example to simulate the entrance to a tunnel, it is possible and useful to take into account the real elevation of the topographic surface in the model.

In terms of model thickness, in the case of energy segments it is possible to reproduce a certain number of tunnel rings equipped for heat exchange sufficient to also take into account the reciprocal thermal influence between the rings in the case of groundwater flow not perpendicular to the axis of the tunnel. If, on the other hand, it is a tunnel built using traditional methods, the same care must be taken with regard to the width of the activated module.

To build the model it is necessary to reproduce the correct geometry of the case study, including the arrangement and geometry of the geothermal circuit as well as the hydrogeological and thermal conditions of the subsoil. As for other energy geostructures, it is possible to neglect the thermal diffusion in the thickness of the exchanger tubes to save computational time and to simplify the analysis.

Any groundwater flow present, linked to porous media with high permeability and considerable hydraulic gradients, and the consequent advective-convective heat exchange can be reproduced by imposing a difference in hydraulic head as a boundary condition, on the lateral edges of the model, which therefore determines a flow in the direction corresponding to that observed in situ, as a function of the hydraulic conductivity of the soil.

As thermal boundary conditions at the upper boundary, in the case of sufficiently deep tunnels (at least 10–15 m of cover) it is possible to neglect the influence of the external air temperature at ground level, directly imposing the undisturbed ground temperature. Alternatively, it will be necessary to have the most recent external temperature variation data, to be imposed as a Dirichlet or Cauchy boundary condition on the surface, in order to obtain a temperature profile that varies with depth (heterothermic zone). For the lower domain boundary, the temperature is usually fixed at the undisturbed ground value and the geothermal heat flux is neglected. However, in certain cases, such as those already mentioned in Sect. 3.2.2, a geothermal heat flow can be imposed (Neumann type boundary condition). The lateral boundaries of the domain are usually considered adiabatic.

With reference to the boundary conditions to be imposed on the intrados of the tunnel, the choice is more complex. Typically, the internal air is not simulated explicitly (i.e. there is no mesh inside the tunnel), but there are situations in which it can be important to take this into account, such as for example in the case of the study of alpine tunnels where the objective may also be to control the internal temperature. The choice is therefore left to the designer. A convective flow is often considered on the intrados by imposing a Cauchy boundary condition, defining the internal air temperature history and a conductance parameter which can vary between 0 and 30 W/(m^2K) and which basically depends by the speed of the air flow and the roughness of the tunnel walls as better illustrated in Sect. 3.2.2. However, it is also possible to evaluate the application of a Dirichlet boundary condition, even if it is considered less likely.

As regards the initial thermal conditions in the case of sufficiently deep tunnels, it is possible to assume a constant temperature in the domain and equal to the undisturbed ground temperature. Otherwise, it is good practice to impose an initial temperature that varies with depth and equal to that measured on site in several points of the domain before the geothermal activation of the work or to reconstruct the thermal profile through a multi-year analysis as already described for the energy support structures.

An example of geometry and boundary conditions applied is shown in Fig. 3.6.

Modelling usually consists of several calculation phases:

- the first phase concerns the hydraulic initialization of the groundwater flow if present and takes place in stationary conditions;
- the second phase concerns the thermal initialization and is of a transient type, with a suitable duration (e.g. monthly if the outside air is not considered or some annual cycles in the opposite case). This phase provides the initial thermal state due to the ground temperature, any atmospheric temperature and temperature variations in the tunnel;
- in the third phase the thermal activation of the system is simulated. The initial thermo-hydraulic state is that obtained at the end of the two phases mentioned above in the season corresponding to that of thermal activation. The analysis is carried out by setting the temperature and flow rate at the inlet of the exchanger tubes and calculating the outlet temperature. It will depend on the heat exchange

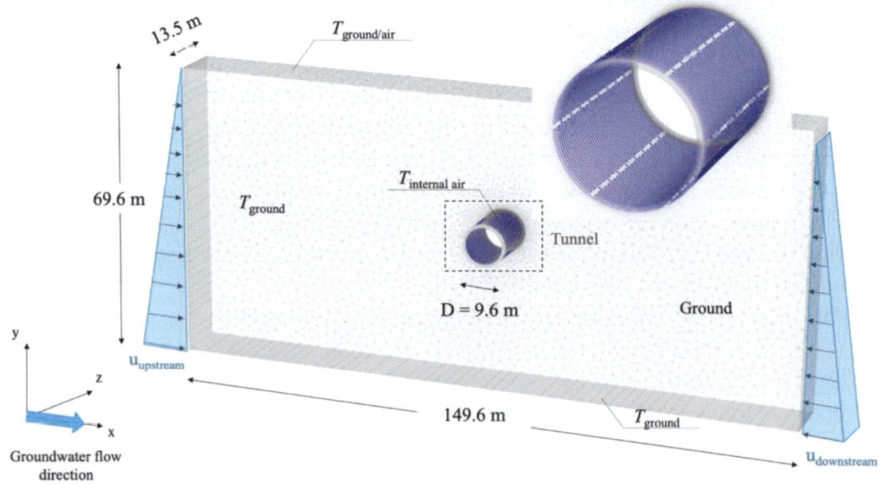

Fig. 3.6 3D view of the numerical model of an energy tunnel with indication of the hydraulic and thermal boundary conditions

that is numerically reproduced during the analysis. By varying the inlet temperature to the exchanger pipes, the analysis can allow to simulate the average behaviour in the various activation periods of the system. For example, by setting an inlet temperature between 24 and 28 °C, summer operation can be simulated, when the excess heat is dispersed into the ground. On the other hand, by imposing an inlet temperature between 3 and 6 °C, it is possible to simulate winter behaviour, extracting heat from the ground. Finally, by cyclically varying the inlet temperature, based on the annual operating modes, it is possible to simulate the seasonal behaviour of the system and evaluate the long-term effects.

Typically, the analyses are performed by reproducing the functioning of the geothermal plant for a certain number of years, to analyse not only the short-term effects, but also the long-term ones.

The thermal power exchanged can be calculated starting from the temperature difference between the inlet and outlet of the pipes and used to determine the geothermal potential of the stretch of tunnel in question. An example output is shown in Fig. 3.7.

In the literature there are also cases in which the temperature entering the circuit is set starting from the thermal load to be satisfied and from the temperatures obtained at the outlet for each calculation step (Bidarmaghz et al. 2021).

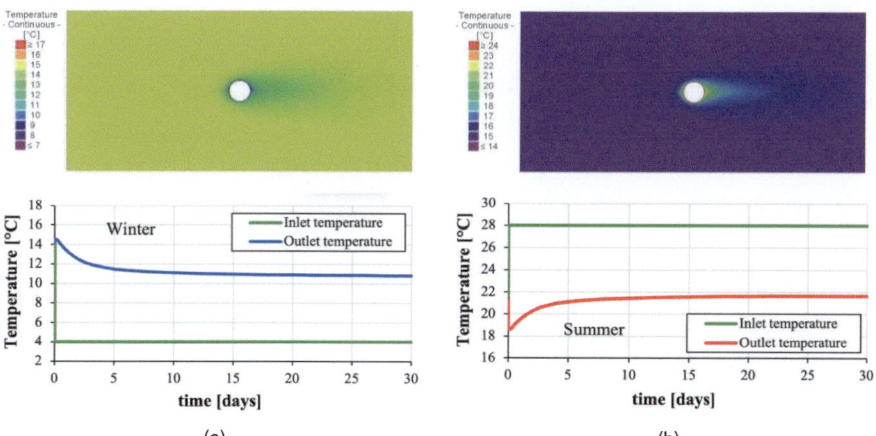

Fig. 3.7 Example of the result of a thermo-hydraulic 3D analysis: ground temperature and temperature calculated at the outlet of the geothermal circuit **a** in winter and **b** in summer

Chapter 4
Thermo-Mechanical Design of Energy Geostructures

Abstract This chapter examines how temperature variations affect the structural and geotechnical behaviour of energy geostructures. Thermal loading induces expansion and contraction, generating additional stresses and strains due to constraints from surrounding soil and structural elements. The chapter explains that while ultimate limit states are generally not critical, serviceability issues such as displacements and stress redistribution must be assessed. It describes analytical methods and numerical modelling to evaluate coupled thermo-mechanical effects. Particular attention is given to piles, walls, and tunnel linings, highlighting interactions within pile groups and structural systems. Overall, the chapter stresses the importance of integrating thermal effects into standard geotechnical design.

Energy Geostructures (EGs) fulfil both a primary structural function and a secondary energetic role. Due to the latter, thermal fluctuations occur within the structure, inducing modifications in the stress and strain state of the structure itself. The purpose of the thermo-mechanical analysis is to verify the occurrence of these side effects and their compatibility with the structural function in accordance with Eurocodes. In particular, it has been observed that thermal variations induce limited effects that, in general, do not compromise safety with respect to the ultimate limit states (ULS) of the structure, but may affect the serviceability limit state (SLS) (Bourne-Webb et al., 2016b).

Thermo-mechanical effects stem from the interaction between the geostructure, which experiences temperature fluctuations, and external constraints imposed by various elements in contact with the structure. These encompass the surrounding ground and other structural components, such as beams, slabs connecting piles, tiebacks, struts, slabs connecting diaphragm walls, bolts, and anchors in tunnel linings.

For example, the temperature variation induced in an energy pile by the heat transfer fluid (initially colder than the pile in winter heating mode and warmer in summer cooling mode) produces a contraction in winter and an expansion in summer. If the pile was free, it would undergo a maximum contraction or expansion, related

M. Barla et al., *Recommendations for the Conceptualisation, Design and Construction of Energy Geostructures*, SpringerBriefs in Applied Sciences and Technology,
https://doi.org/10.1007/978-3-032-30927-3_4

Table 4.1 Summary of tools available for thermo-mechanical design of EGs depending on the design stage

	Feasibility study	**Outlined design**	**Detailed design**
Foundation piles	Conservative estimations of maximum stresses and deformations	Design charts or analytical solutions	Analytical solutions or numerical modelling
Foundation slabs	-	Numerical modelling	Numerical modelling
Walls	-	Numerical modelling	Numerical modelling
Tunnels	-	Numerical modelling	Numerical modelling

only to the temperature variation undergone and to the material's thermal expansion coefficient. However, deformations are not completely free but are partially prevented by the presence of external constraints, such as an embedding in a rigid layer at the base of the pile or in a rigid superstructure at the head, or the lateral soil itself, which frictionally restricts the movements of the pile. If the pile was perfectly constrained at both the tip and head, shortening and elongation deformations would be completely prevented. Consequently, the pile would undergo a maximum variation in internal stresses, tensile in the case of prevented shortening and compressive in the case of prevented elongation. In reality, an intermediate condition between the total absence of restraint and the degree of complete restraint exists, so that the energy pile undergoes both thermally induced strains and thermally induced stresses.

Table 4.1 summarizes the tools available for thermo-mechanical design of EGs depending on the design stage. The different tools are described in the following.

4.1 Energy Piles

Geotechnical design of an energy pile or a group of energy piles is performed on the same basis as conventional piles, however with additional check of thermal stresses developed in the piles and additional settlements of piles due to the temperature variations. Therefore, energy piles must fulfil the requirements of ULS and SLS, as defined in Eurocode 7 standard (EN 1997–1) with consideration of thermal loading in addition to the classical loads.

To obtain the design load the mechanical and thermal actions are taken into account. The permanent and variable actions are identified and the appropriate load combinations with partial safety factors for permanent and variable loads are assigned together with accompanying factors when two or more types of loads are applied simultaneously.

Thermal actions could be neglected in the design of energy pile at the ULS if sufficient ductility and rotation capacity of piles is ensured. Both mechanical and thermal actions are considered when checking requirements of the SLS.

The main requirements that an energy pile must fulfil are:

- The vertical stress variations caused by mechanical and thermal loads must be considered with respect to the entire length of the energy piles, and the most stressed section must be verified.
- The vertical displacement variations caused by mechanical and thermal loads must be considered with respect to the head of the energy piles.

As for conventional piles, ULS design for energy piles includes the following checks:

- Bearing capacity of pile (base and shaft capacity),
- Structural capacity of pile (axial force, bending moment, shear force).

Since one of the main requirements of the ULS is to ensure the bearing capacity of the pile, and this is related to the shear strength of the soil, taking into account the heating and cooling effects, should be considered in the design of the ULS for energy piles.

The main requirements that must be satisfied when designing energy piles are (GSHPA, 2012): pile settlement, differential settlements, axial, bending and shear stresses, negative skin friction, axial and radial pile expansion/contraction/fixity, thermally induced axial stresses, cyclic effect of the thermal actions and temperature at soil-pile interface including daily/seasonal variations.

The SLS should be evaluated for the displacements expected under both externally applied and thermal loads.

The mechanically induced and thermally induced stresses and vertical displacement of the energy piles are estimated with analytical methods such as load transfer method or with Finite Element Method (FEM). The comparison of the load-transfer method and FEM was presented by Loveridge et al. (2020b). In the load-transfer analysis the main disadvantage is the inability to analyse group of piles and the need to define the appropriate head restraint boundary conditions and to take into account the radial stresses. The main issue in FEM is to assign the material model with constitutive relationships that adequately describe the nonlinear response of the soil due to the mechanical and thermal actions (also see Chapter 3). FEM also enables to evaluate the thermo-hydro-mechanical response of the ground surrounding the energy pile and can also evaluate the cyclic effect of the temperature variations with time. One of the main advantages of FEM is the capacity to simulate the energy piles system with time, in an integrated way focusing on both thermo-mechanical and thermal behaviour basing the design on an analysis that captures the performance of the system for 5, 10, 25, 50 years.

Analytical solutions for energy piles usually assume a linear thermoelastic behaviour for the piles, which means that their deformation and thermal properties are directly proportional to the applied loads and temperature changes. End restraints of the pile have a significant effect on the mobilization of shaft resistance and, consequently, on the displacement of the pile and the stresses generated in the pile. In real projects, where energy piles play the role of a building foundation, the head restraints are never free. The degree of restraint at the head is significantly influenced by the loads transmitted from the building to each of the piles and the type of structure connecting the piles together, forming the foundation system. The base restraint conditions however are directly impacted by the mechanical properties of the material (soil or rock) found at the base of the pile. Therefore, a combination between thermal induced displacements and thermal induced stressed is a realistic assumption regarding the thermo-mechanical behaviour of energy piles.

The vertical deformation of a single energy pile and the resulting stress development, induced by a combination of mechanical and thermal loads, are contingent upon several crucial parameters. These include:

- The stiffness of the pile and the surrounding soil.
- The geometric characteristics of the pile.
- The end-restraints conditions.
- The thermal expansion potential of both the pile and the soil.
- The duration of geothermal operation of the energy pile (i.e., the time involved).
- The rate at which the loads are applied.

When a pile is heated, the concrete tends to expand; when cooled, it contracts. If the pile head, base, or shaft friction constrains that movement, part of the thermal strain is converted into stress. Under heating, compression generally increases, while cooling can reduce compression and even generate tension in extreme cases.

In practical operation, an energy pile experiences temperature fluctuations linked to seasonal heat-exchange cycles. The temperature variations recorded in various studies along the pile shaft typically range from + 30 °C during heating to − 15 °C during cooling. When group of piles are considered (scenario corresponding to real cases) their individual thermal influence zones may overlap depending on their positioning (piles spacing) which may cause local ground temperatures to merge. As an effect the effective ΔT around each pile may increase up to 30–40%.

Figure 4.1 explains the principle of thermal effects within the energy piles and their influence on the thermo-mechanical behaviour of the piles. An energy pile subjected to mechanical loading develops axial compressive strains and stresses governed by the applied structural loads and by the interaction with the surrounding soil through shaft resistance. When the pile is thermally loaded, a temperature increase induces axial expansion, while a temperature decrease induces contraction; however, the thermal deformation is partially restrained by the surrounding ground and by boundary conditions at the pile head and toe. As a consequence, heating generates expansion within the pile, whereas cooling may induce contraction. Under combined thermo-mechanical loading, the pile response results from the superposition of mechanically induced and thermally induced strains and stresses, leading to a coupled behaviour characterised by redistribution of axial force and shear stress along the pile shaft. Depending on the magnitude, sign and sequence of thermal and mechanical actions, thermal expansion may either amplify or partially counteract mechanical stresses, shifting the location of the Null Point and modifying load transfer mechanisms.

Thermal expansion and contraction of the pile length produce measurable axial strain changes (Table 4.2). In existing data from single-pile tests vertical strains may

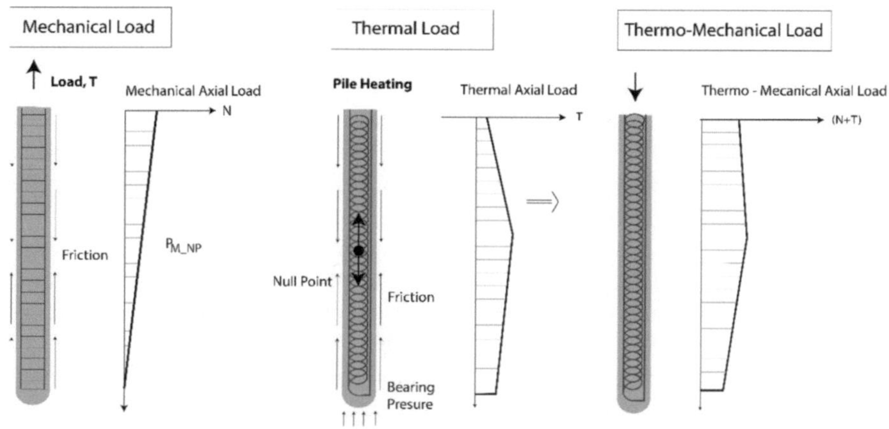

Fig. 4.1 Schematic illustration of thermo-mechanical behaviour of energy piles

reach up to approx. 250–300 με under heating of ΔT ≈ 20–30 °C. The strain distribution is roughly uniform along most of the pile length. Radial expansion accompanies axial elongation due to the Poisson effect. The typical radial strain in concrete piles is about one third of the vertical strain. This results in a small change in pile diameter, on the order of 0.1–0.2 mm for a 1 m pile diameter. However, while these deformations are minor structurally, they may influence the contact and interface shear stress between the pile and soil, thereby modifying load transfer mechanisms. In the case of energy piles groups, their positioning may determine a superposition of temperature fields which slightly increases the average thermal strain, particularly in the upper portions of piles. However, usually the thermally induced vertical and radial strains remain elastic in nature.

Restraints at the pile head, toe, and along the shaft convert blocked thermal strains into stresses. Thermally Induced Vertical Stress may reach values in the order of 1–6 MPa of stress depending on the restraints conditions. Field evidence shows

Table 4.2 Ranges of variations for energy piles

Parameter	Single Pile (Typical Range)	Group Effect (Relative Change)
Temperature variation (ΔT)	− 15 °C → + 30 °C	+ 10–40% higher near center piles
Vertical strain ($\Delta\varepsilon^z$)	± 100–250 με	+ 15–25%
Radial strain ($\Delta\varepsilon^r$)	± 30–80 με	+ 5–10%
Vertical stress ($\Delta\sigma^z$)	+ 40 → + 360 kPa / °C (heating) − 15 → − 180 kPa / °C (cooling)	up to + 200%
Head displacement	± 2–6 mm (free head) ± 1–3 mm (restrained)	× 1.5–2 increase; uneven across group

actual increments between 40–360 kPa per °C during heating cycles and increments between 15 -180 kPa per °C for cooling. Thus, over a 20 °C cycle, thermal loading can alter axial stress by 4–7 MPa, comparable to mechanical design loads. *When* multiple piles are simultaneously heated axial stresses tend to increase due to mutual restraint and increased soil confinement. Recorded and simulated results indicate stress magnitudes can rise by up to 100–200% compared with isolated piles especially when pile spacing is smaller than 4 to 6 diameters of the piles.

Temperature changes also affect vertical movement at the pile head. In heating phases, piles may develop head displacement in the range of mm ($\approx$ 1–3 mm). During cooling, contraction combines with structural load and produce additional settlements of comparable magnitude, typically 2–5 mm that depend on the ΔT. Usually, the displacements depend on the pile's boundary conditions, mechanical load, and soil compressibility (ex. non/normal-consolidated soils tend to settle more under heating, higher impact on cohesive soils compared to noncohesive). Pile groups usually generate larger cumulative ground deformation with vertical displacements at the group cap that can be 1–2 times higher than those of a single pile. Also differential heave or settlement may appear if piles operate under uneven thermal loads.

The thermo-mechanical assessment and design of energy piles should rely on methods that can accurately represent the coupled interaction between temperature, stress, and displacement within both the pile and surrounding soil.

4.2 Load-Transfer Method

Load-transfer method uses the measured relationship between the pile resistance and pile movement at various points along the pile (Knellwolf et al., 2011). Therefore, the curves relating the ratio of the adhesion and the soil shear strength to the pile movement are needed in analysis and are obtained from the field and laboratory tests (Poulos, 1989),. The load transfer effect based on the assumption that the temperature changes are uniform over the lengths of the pile was presented by Bourne-Webb et al. (2009). Since the settlement of the pile head depends on the boundary conditions, the following options for pile body restraint should be considered:

- Floating pile (no pile head nor pile base restraint),
- Pile head fixed along with no pile base restraint,
- Partly fixed pile head along with no pile base restraint,
- Free pile head along with finitely rigidity of the pile base,
- Pile head fixed along with finitely rigidity of the pile base,
- Partly fixed pile head along with finitely rigidity of the pile base,
- Free pile head along with infinitely rigidity of the pile base,
- Pile head fixed along with infinitely rigidity of the pile base, and
- Partly fixed pile head along with infinitely rigidity of the pile base.

Figure 4.2 illustrates conceptual variations between the degree of restriction at the head of the pile, top load and corresponding settlements for different levels of

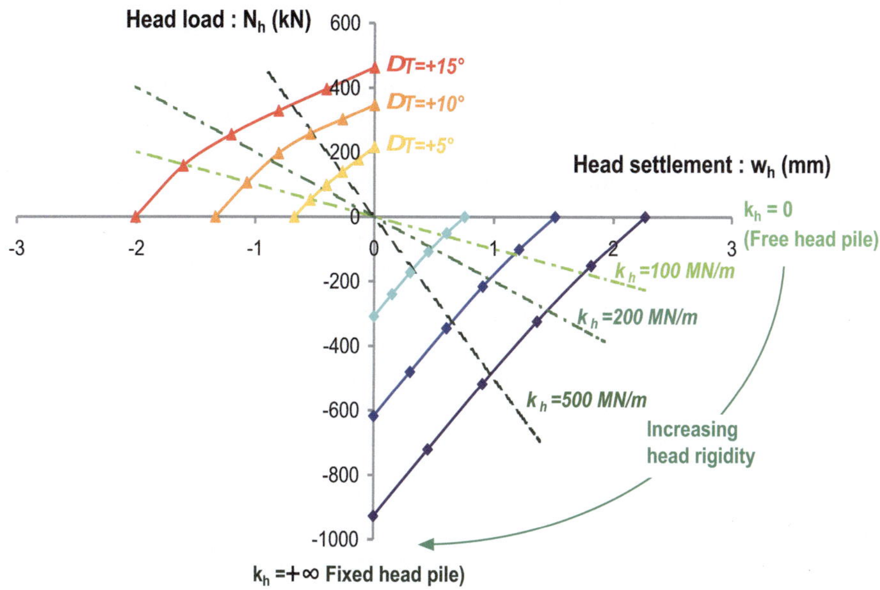

Fig. 4.2 Typical diagram of head settlement and head load variation due to thermal loading for different head rigidity levels (Burlon et al., 2013)

head rigidity k_h. The figure shows how the settlements evolve at different variations of top load and various degrees of freedom of the pile. As a result, the higher the top load the higher the settlements for lower degree of freedom. The higher the restraint conditions at the tip of the pile which can translate into stiffer material/soil/rock at the bottom, the higher the stress developed in the pile. Overall, Fig. 4.2 captures the essential thermo-mechanical coupling governing energy pile behaviour—linking restraint conditions of the pile, settlement evolution with load variation of the pile.

4.3 Finite Element Method Considering Thermal Cyclic Effects

The load-transfer curves method is a potent tool in design practice, enabling a preliminary and accurate sizing of energy piles. However, this method falls short in considering several vital aspects that could significantly influence behaviour, particularly the constitutive behaviour of the soil and the complexities of heat and water diffusion. For a comprehensive account of these physical aspects, the numerical methods (e.g. FEM, FDM) proves more suitable.

The finite element method (FEM) may provide a full three-dimensional representation of the pile–soil system and explicitly models heat transfer and mechanical equilibrium. FEM captures the transient evolution of temperature fields, and the resulting stress redistribution both within the pile and in the surrounding ground. Compared with the load-transfer method, finite element analysis offers greater accuracy and insight into complex boundary conditions, group interactions, and non-uniform heating or cooling scenarios. In practice, the load-transfer approach is often used for preliminary and routine design, while FEM simulations are reserved for final verification, optimization, and research-grade analysis of energy geostructure performance.

The results obtained from the more comprehensive 3D simulation of an energy pile group offer valuable insights into the understanding of pile interactions and analysing the overall foundation response under mechanical loading and thermal cyclic stresses. Typically, the energy piles, which are made from reinforced concrete, are assumed to behave thermo-elastically in the numerical simulations. Several constitutive models have been developed for modelling the stress–strain behaviour of soils, but few consider the thermal cyclic aspect. One such constitutive model for clay is the ACMEG-T model (Modaressi & Laloui, 1997).

In general, pile design assumes axial loads, bending moments and shear forces. Pile design should consider the slenderness of the pile to determine the appropriate moments. Piles are also designed and checked for biaxial moments if they occur. For the limit state STR, which deals with the failure of the pile, a general method for designing piles is as follows:

- determine the design life of pile,
- assess actions on the pile,
- determine which combinations of actions apply,
- assess durability requirements and determine the concrete strength,
- check cover requirements for appropriate durability and bond,
- analyse pile for critical combinations of moments and axial forces,
- check slenderness and determine bending moments,
- determine area of reinforcement required and
- check spacing of bars and links.

In practice, several of the above steps can be combined, and these steps are applicable to both conventional pile and energy pile design.

The load-bearing capacity of a pile group could be significantly greater than that of a single pile foundation. However, this group capacity is heavily influenced by the interaction between the piles and the soil, which is contingent upon factors such as the group configuration, group size, and spacing between individual piles. In practice, a pile cap is employed to connect the heads of the piles directly with the ground, creating what is known as a piled raft system. This system enhances the bearing capacity due to the presence of the cap, but introduces a complex interaction among the piles, the ground, and the foundation slab. In situations where thermo-active piles are employed for heating/cooling purposes in a building, not all the foundation piles need to serve as energy piles. The number of such piles required depends on the building's energy demands and the ground's energy source capacity. The distinct responses of activated and non-activated piles can lead to different behaviour than predicted for a single pile, as loads are redistributed through the foundation slab.

The partial use of the foundation as thermo-active elements can introduce significant additional stresses due to the stiffness of the overlying structure Dupray et al. (2014). The extent of these additional stresses experienced by non-thermal piles depends on the stiffness of the foundation slab. In contrast to load redistribution caused by mechanical loading, higher stiffness should be considered unfavourable as it leads to more extensive transfer of thermally induced displacements to other piles. Rotta Loria and Laloui (2016) presented design charts for the displacement behaviour of energy pile groups using the interaction factor method.

4.4 Energy Walls and Slabs

As for the case of thermo-active piles, the thermal effects represent additional contributions to the mechanical loads. However, with respect to energy piles, energy walls are characterized by a more complex geometry, non-axisymmetric, with structural constraints such as anchors and base slabs, one side partially exposed (to air/water) and not in contact with the soil, mainly horizontal loads, eccentric thermal loads and a bending type of behaviour. All these aspects make thermal effects less predictable compared to piles. Additionally, it is not possible to directly apply the concepts described for energy piles, therefore energy walls need to be addressed ad hoc. Similar conditions apply to energy slabs even though geometry is different and they are generally active at a shallower depth.

Thermo-mechanical effects in energy walls due to thermal loading are represented by possible changes in horizontal and vertical displacements of the soil and of the structure (and, therefore, of settlements of buildings at the surface) and by changes in structural internal forces and soil stresses. These effects are highly dependent on soil initial temperature, on temperature changes with time at the surface, on temperature changes related to the operation of the geothermal system and on the user energy demand, on the thermal boundary conditions (in particular at the excavation side), on the constraints due to the soil at the side and at the base, on structural constraints (actions from connected anchors or slabs), on thermal and mechanical properties (such as stiffness) of the involved materials, on mechanical loads (Habert & Burlon, 2015; Bourne-Webb et al., 2016a; Sterpi et al., 2017; Barla et al., 2020a).

In the literature thermo-mechanical effects are usually studied resorting to 2D or 3D finite element or finite difference numerical models with coupled thermo-mechanical or thermo-hydro-mechanical approach. Currently, simplified analytical or empirical methods are not available. The analysis of energy walls, that often have complex geometry and boundary conditions, can profit considerably from the accurate modelling offered by these methods. Figure 4.3 shows a couple of examples of numerical mesh discretization for the simulation of thermo-mechanical behaviour of an energy diaphragm wall. Beyond the traditional mechanical parameters, it is necessary to define thermal conductivity, specific heat and the thermal expansion coefficient for all materials involved in the model. The simplest case is to consider an elastoplastic constitutive model for the soil (such as Mohr–Coulomb, Cam-Clay)

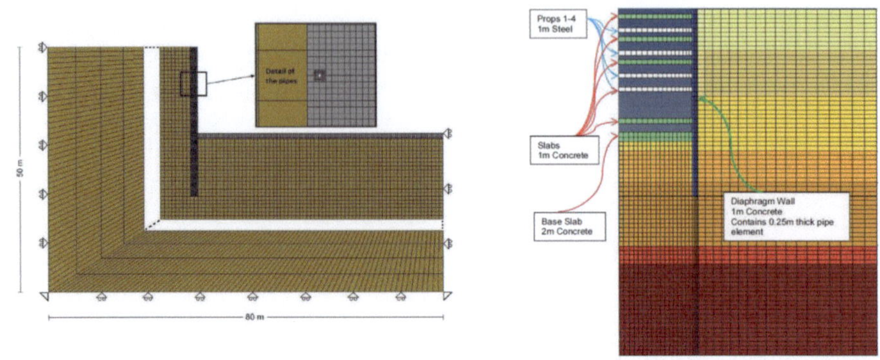

(a) (b)

Fig. 4.3 Examples of (a) FDM and (b) FEM model for thermo-mechanical analysis of energy diaphragm walls (Barla et al. 2020a; Soga et al., 2015)

considering in the elastic law an additional component depending on temperature in an isotropic and linear way (thermal variations only induce reversible volume changes) and a thermo-elastic behaviour for the structural material (such as concrete). If this approach can be accepted for coarse-grained soils, it's worth specifying that for fine-grained soils more complex constitutive models exist in literature (such as the dependence of the hardening law from temperature).

The stages usually considered in thermo-mechanical numerical analyses of energy diaphragm walls include the geostatic initialization of the domain and the initialization of the stress–strain state of the diaphragm wall by reproducing the excavation and construction process, since the final stress–strain state is a function of it. This stage is time independent. If needed, the domain is initialized from a hydraulic point of view by means of appropriate hydraulic boundary conditions (hydraulic head or pore pressure) so as to reproduce the static regime or steady-state seepage. Usually a thermo-mechanical analysis stage, without thermal activation, follows, to obtain the initial temperature field. During this stage the effect of thermal environmental variations on the geostructure's stresses and strains can be analysed. Then thermal activation is simulated during which it is evaluated how the stress–strain state previously calculated varies in time and in space. At the end of the analysis, it is possible to evaluate earth pressures, displacements and structural forces at different time frames.

For instance, according to a study by Sterpi et al. (2017), whose results can be observed in Fig. 4.4, the horizontal earth pressure changes, influenced by materials thermal expansion, are practically negligible with respect to the undisturbed natural conditions and within 10%. Similarly to piles, a thermal contraction of the structure is observed in winter and a thermal elongation in summer in the order of 13% at the head (about 2 mm) with respect to undisturbed conditions. The increase in horizontal displacements is in the order of 8% (7 mm) and oriented towards the excavation side. Therefore, in this case, as well as in other cases from literature, thermal effects appear acceptable in terms of geotechnical safety.

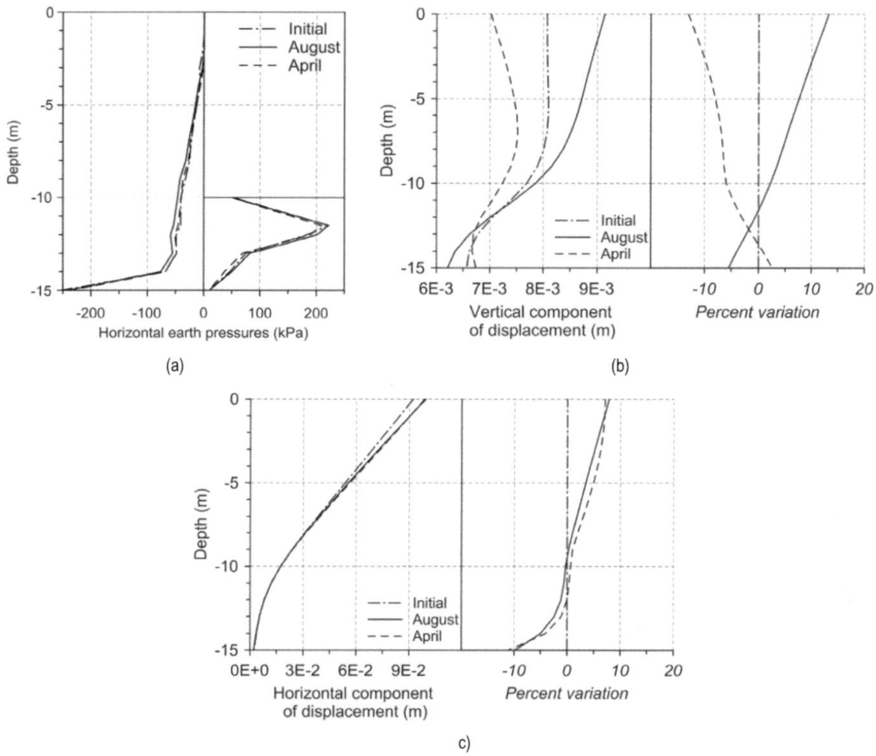

Fig. 4.4 (a) Horizontal earth pressure on the structure and (b) vertical and (c) horizontal components of the displacement (Sterpi et al., 2017)

However, the small changes in displacements imply thermally induced stresses and, hence, internal forces, that cannot be completely neglected.

In Fig. 4.5, in addition to alternating compression and tension conditions according to the season, three other fundamental aspects emerge.

- The first one concerns the maximization of the axial force at the beginning of the heating or cooling season, since, although maximum temperature variations occur at the end of the season (April and August), it is at the beginning that there are maximum temperature gradients between the structure and ground (November and June respectively). The latter, therefore, not yet affected by the heat exchange process, tends to oppose to heat-induced deformations.
- The second aspect concerns the different behaviours of different adjacent cross sections since there is a temperature gradient in the longitudinal direction and the mutual interaction among the cross sections prevent each other from thermal elongation or contraction. An increase in compression is observed in summer and a decrease in compression, which can lead to traction, in winter in the section closest to the heat exchangers, and an opposite effect in the sections less affected by the temperature variation. For this reason, a three-dimensional model is recommended, which is more accurate than a conventional plane strain model.

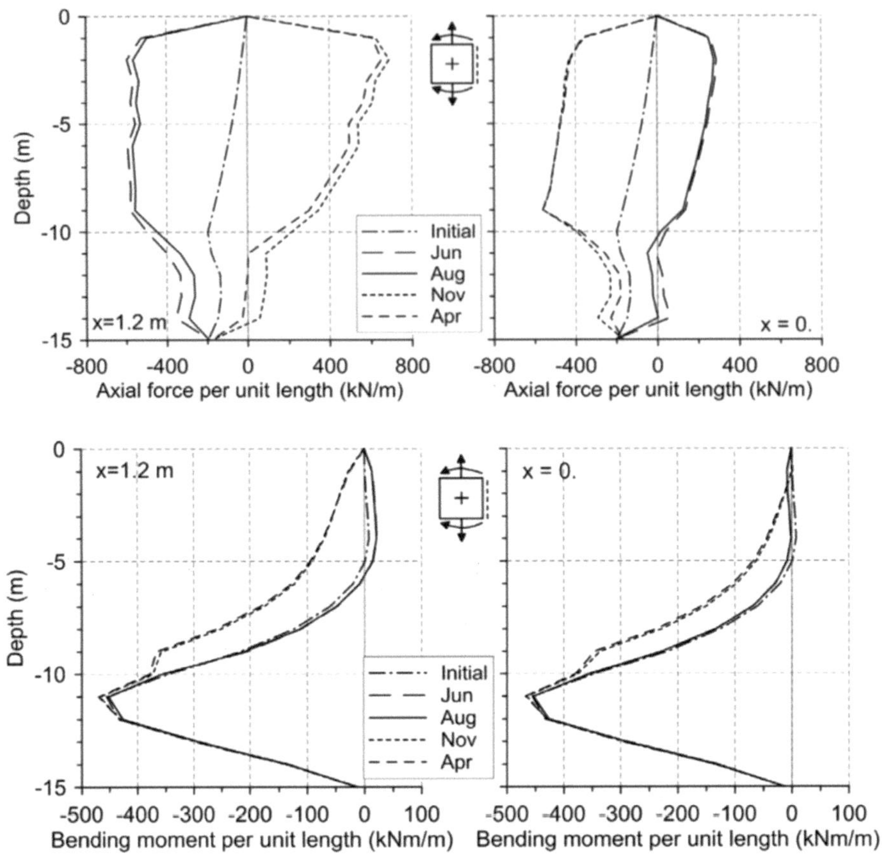

Fig. 4.5 Axial force and bending moment at different time frames in two different vertical cross sections (Sterpi et al., 2017)

- Finally, even though some change in bending moments will occur, in the embedded part of the wall, these changes are negligible. The presence of the base slab and of the ground on both sides of the immersed part contribute to reducing the bending moment in the deep section of the structure.

Similar results also emerge from Barla et al. (2020a) (Fig. 4.6) where the horizontal displacement varies between –4% in winter and + 16% in summer with respect to the environmental thermal oscillations. The maximum variation of the bending moment is obtained in winter (+17%) when the displacements are negligible. The authors conclude that the stress variations are largely included within the strength limits of the structure.

The effect of heating and cooling cycles on the short- and long-term structural behaviour of an energy retaining structure was studied by Soga et al. (2015) for the case of the Tottenham Court Road station of the Crossrail in London, in clayey soil. From the numerical results, the most relevant effects are the deformations induced by a thermal differential along the thickness, the variations in active and passive earth

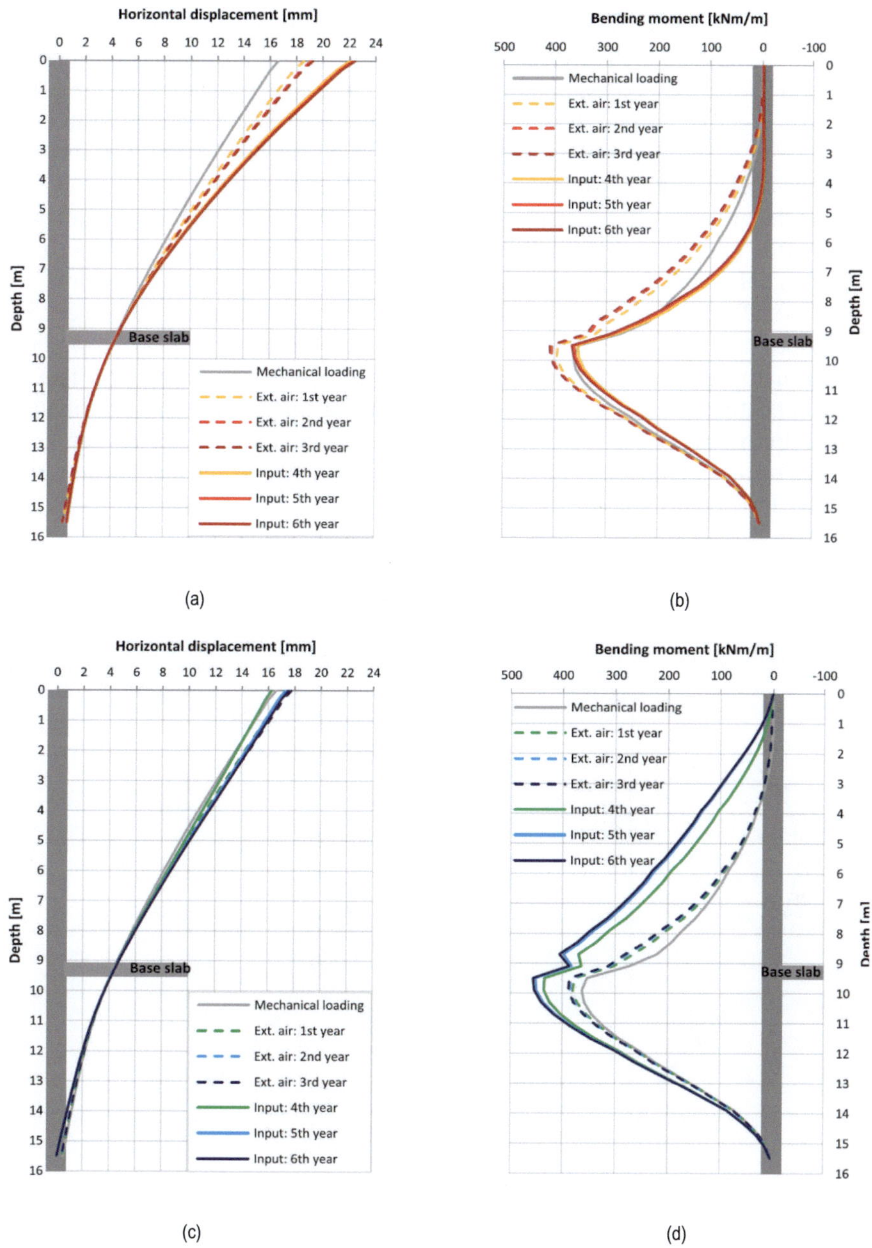

Fig. 4.6 Horizontal displacements and bending moment solicitations at different time frames during summer (a), (b) and winter season (c), (d) (Barla et al., 2020a)

pressure and the shrinkage or expansion of the soil due to a long-term thermal drift. However, the numerical prediction indicated minor variations of earth pressures and minor soil shrinkage/expansion due to a ground thermal drift in the long term. Due to a thermal gradient in the wall cross-section, concrete strains induce an increase in calculated bending moments, but usually the moment distribution remains well within the design failure envelope adopted for the wall design.

Since the ground-wall interface resistance is not a key factor in the wall structural function, the possible detrimental effect induced by cyclic thermal loading can be neglected, unless the diaphragm wall is used also with a load bearing function.

Additional studies (Habert & Burlon, 2015; Bourne-Webb et al., 2016a) confirm that temperature variations have a limited effect on the behaviour of the structure which usually still complies with the requirements to ensure safety and that the effects of heating and cooling operations are minimal compared to those attributable to natural climatic fluctuations. Most of the thermo-mechanical effects are hence produced by the natural temperature variations compared to which thermal activation adds very little contribution (Fig. 4.7).

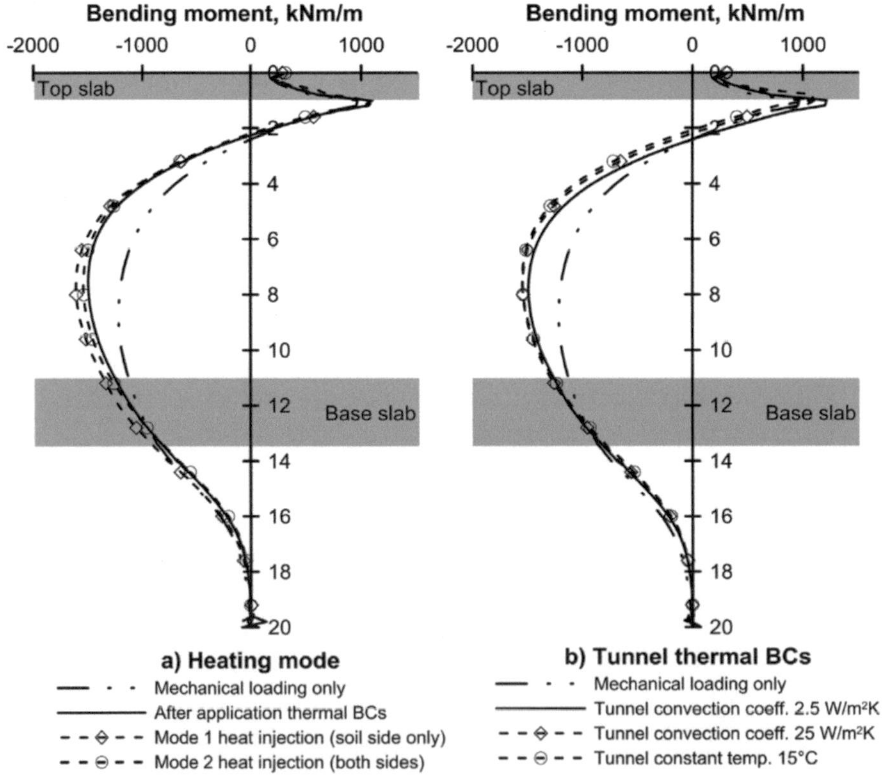

Fig. 4.7 Example of the variation of bending moment along an energy diaphragm wall, for different heating modes and thermal conditions at the excavation side (Bourne-Webb et al., 2016a)

To conclude, the thermally induced mechanical effects, observed partly as additional displacements and partly as variations of internal actions (axial forces and bending moments), are limited and do not seem to threaten the structural and geotechnical safety of energy walls as they are most likely below the strength limits of the structure. However, additional stresses may arise that require careful evaluation during the design phase to anticipate possible unexpected overload situations. The structural conditions and the constraints are of particular importance; therefore, it is essential to correctly simulate the construction sequence and the connection structures in the numerical analysis in order to obtain reliable results.

More recently, analytical solutions have also been presented in the literature (see Rotta Loria et al., 2020), although their use in practice has not yet widespread and whose application is limited by the assumption of a linear temperature gradient across the structural element thickness.

4.5 Energy Tunnels

As for piles and diaphragm walls, also for tunnels thermal activation of the lining can lead to side structural effects that need to be adequately evaluated, both in terms of strains and in terms of stresses due to prevented strains. In particular, temperature changes generated by heat exchange between the heat carrier fluid and the surrounding soil/rock mass and/or the inner tunnel environment could involve:

- thermal expansion/contraction cycles of the geological formation hosting the tunnel with subsequent change in the load acting on the lining;
- thermal expansion/contraction cycles of the lining with subsequent change in axial force and bending moment of the lining;
- change in the effective stress state of the soil also due to a stress redistribution induced by the above-mentioned load and internal actions changes.

Such possible phenomena induced by temperature variation are schematically illustrated in Fig. 4.8.

Based on the above, temperature distribution in an energy tunnel needs being estimated accurately to determine the change in induced strains, the additional thermal stresses, and compare the thermally induced actions with the resisting ones. To this aim, at present no simplified analysis methods exist, but it is possible to refer to previous experiences or to perform thermo-mechanical analyses to quantify the lining stress–strain conditions.

The calculation steps required to carry out a numerical simulation include (i) the geostatic initialisation of the domain and the initialisation of the stress–strain state of the tunnel lining under operating conditions, which requires simulation of the excavation sequence and construction phases; (ii) possible hydraulic initialisation of the domain by means of specific hydraulic boundary conditions (hydraulic head or porewater pressure), reproducing the hydrostatic or stationary seepage regime; (iii) thermal initialisation of the domain, through the application of an undisturbed

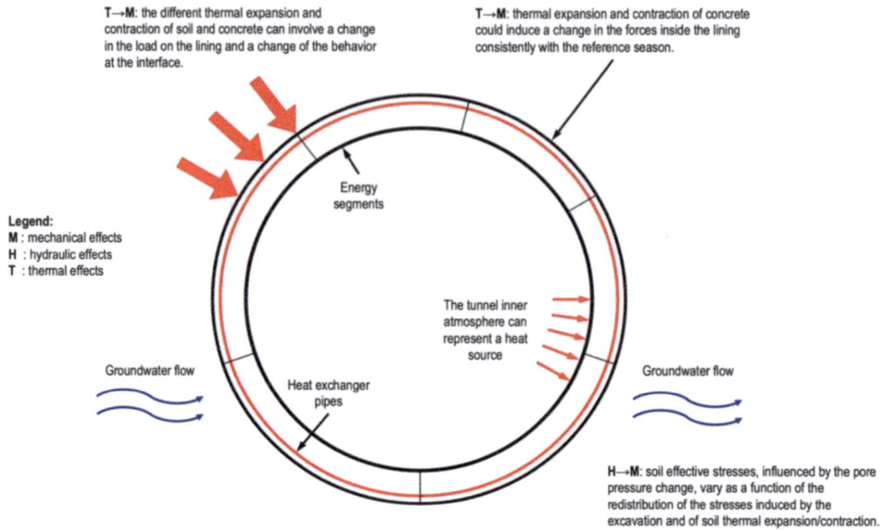

Fig. 4.8 Main issues involved in the geotechnical and structural design of energy tunnels (modified from Soga & Rui, 2016 and Ma et al., 2022)

temperature field or cyclic seasonal thermal solicitations, followed by a transient thermal or thermo-hydraulic analysis, until the final cyclic stationary condition is reached; (iv) activation of the geothermal system by applying a thermal solicitation to the lining. In phase (iii), the effect of environmental thermal variations on stresses and strains in the lining can be analysed, whereas in phase (iv) it is possible to evaluate the effect of thermal activation on stresses and strains in the lining at different timeframes, also with respect to those assessed in (iii).

An example of the thermo-mechanical implications on the lining induced by heat exchange with the surrounding environment is that of the experimental real-scale, two-ring energy tunnel prototype installed in Turin (Italy). The prototype was built during the construction of the southern extension of Metro Line 1 in an alluvial soil, consisting of loose to weakly cemented sands and gravels (Barla et al., 2019; Insana et al., 2020; Insana, 2020). The two energy rings were instrumented with a specifically designed monitoring system to observe stresses, strains and temperatures in the lining. The tests performed in cooling and heating mode made it possible to get insights into the behaviour of energy tunnels. The effects produced by an experimental test in winter mode (heating the user) and in summer mode (cooling the user) are shown in Fig. 4.9. The figure depicts the primary circuit inlet and outlet temperatures, the temperature in the tunnel, the temperatures in the lining, thermally induced strains and thermally induced stresses. During the winter test, the circulation of the heat carrier fluid reduces the temperature of the lining by 4–6 °C, which results in a hoop strain of 50 $\mu\varepsilon$ in contraction and a decrease in radial and hoop compression between 0.2 and 0.5 MPa. These stresses are due to the share of strains prevented due to the constraint conditions to which the lining is subject. During the summer test, the opposite effect occurs, with a 5–7 °C increase in lining temperature resulting in a hoop expansion of 60 με and an increase in radial and circumferential compression between 2 and 3 MPa. The effect is very similar to that due to variations in internal air temperature.

During all the tests carried out, temperature variations in the lining were in the range -7.5–8.3 °C.

About the strain state, thermally induced strains of about 10 με/°C and a substantially elastic behaviour are generally observed. Experimental evidence and numerical studies in the literature confirm that thermal activation of tunnel linings causes limited strain changes and negligible settlements (studies indicate the possibility of millimetre-order displacements under typical operating conditions).

About the stress state, thermally induced stresses in the order of the MPa, between -0.8 and 3.6 MPa are observed. In all the tests carried out, no stresses in the longitudinal direction were ever detected. In the radial and hoop directions, lining stress variations as a function of the temperature change are in the order of 0.1 MPa/°C in heating mode and between 0.1 and 0.5 MPa/°C in cooling mode. These results have confirmed previous numerical studies such as the one described by Nicholson et al. (2013, 2014), concerning the design of an energy segmental lining of a tunnel in England, and in Barla & Di Donna (2018). Additional studies then further validated the empirical evidence above, by also showing how the thermal stresses decrease with decreasing soil stiffness (Ma et al., 2022; Rotta Loria et al., 2022).

Given the limited tensile strength of concrete with respect to the compressive strength, the winter operation must be carefully analysed as, due to the reduction of the compressive stresses, additional tensile stresses could arise. This aspect is particularly worthy of attention in the case of shallow tunnels (for example in urban areas) which are more frequently characterized by low compressive stresses due to the limited cover. In summer (cooling mode) the additional compressive stresses could instead alleviate possible tensile stresses.

About thermally induced actions in the lining, in addition to those due to the excavation and purely mechanical, the effect of thermal activation is an increase in normal compressive axial force in summer and a decrease in winter. Variations in bending moment and shear force resulted to be negligible. The structural verifications at ULS and SLS, carried out considering the thermal action as a variable load, were all satisfied without modifications either to the reinforcement area or to the dimensions of the cross section already defined based on the traditional structural design (i.e. purely mechanical). This consideration can streamline the design and verification of energy tunnels, especially in the early stages of the project. Therefore, during the feasibility study conventional checks could be performed against mechanical loads only.

Further studies (Ma et al., 2022; Rotta Loria et al., 2022; Gawecka et al., 2021) confirm the above and show that:

- no additional plastic zones emerge in the soil following thermal activation compared to those obtained at the end of excavation;

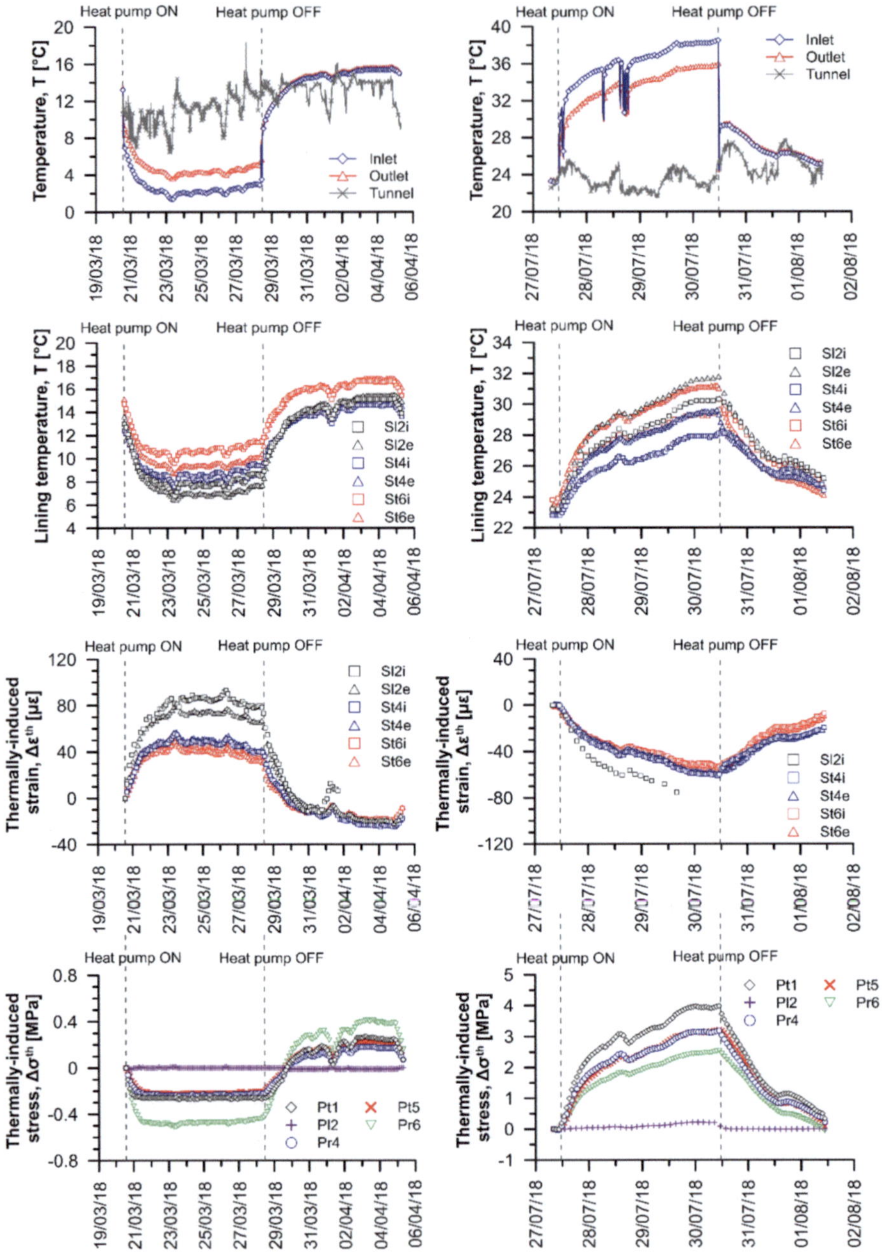

Fig. 4.9 Example of thermo-mechanical effects induced by thermal activation on the lining: winter mode test (left) and summer mode test (right) (Insana, 2020; Insana et al., 2020)

- the asymmetric temperature distribution induced by the possible presence of groundwater flow leads to asymmetric displacement distributions, with maximum variations of 1.5 mm compared to the end of excavation, and to a difference in circumferential stress between the right and left springlines, however included within 1.5 MPa;
- as the groundwater flow rate increases, the circumferential stress variation increases, in a non-linear way, especially from 0 to 0.5 m/d;
- the impact of the thermal conductivity and thermal capacity of the soil on the circumferential stress variation is negligible:
- the role of the thermal expansion coefficient of the soil on the mechanical response of the tunnel appears to be relevant, especially in the presence of groundwater. The stresses in the lining seem to vary linearly with it;
- in highly permeable soils no pore pressures are generated or these dissipate very quickly, without consequences on the structural behavior;
- based on the internal air velocity and the temperature differences between internal air, heat carrier fluid and soil, the thermally induced stresses can also be compressive in the heat extraction phase and also tensile in the heat injection phase;
- the stress state in the lining is highly non-uniform and can include both tension and compression at the same time.

To conclude, the results described above show that although effects induced by thermal loads on tunnels occur (and are measurable), stress and strain changes are, in general, pretty limited and do not threaten the structural integrity or functionality of the lining and are entirely acceptable. Further studies are needed in particular deposits, such as contracting clays following an increase in temperature.

To deepen the thermo-mechanical aspects and study the impact of temperature variations, it is possible, to build an ad hoc numerical model with a coupled thermo-mechanical or thermo-hydro-mechanical formulation, which may also be two-dimensional. During the investigation phase, it will be necessary to collect the mechanical and thermal properties of the materials involved, site stress history, the flow and temperature characteristics of the groundwater, as well as to deepen the knowledge of the aerothermal characteristics of the tunnel (speed and temperature of the internal airflow) and the coefficients of thermal expansion of the materials involved (soil, concrete, grout). The structural verifications at ULS and SLS must take into account the effect induced by the heat exchange in accordance with current regulations, considering the thermal load as a variable load.

If the magnitude of the thermal loading is greater than that considered in the aforementioned studies, for example in the case of heat storage from a solar thermal system, attention must be paid in the design process. Even in the case of twin or neighbouring tunnels, attention must be paid to interaction effects.

Chapter 5
Environmental Studies, Interactions and Interferences

Abstract Chapter 5 focuses on environmental impacts and interactions between energy geostructures and surrounding systems. It explains thermal interference, which occurs when heat transfer from one installation affects another, especially in dense urban environments. The chapter discusses heat transport mechanisms (conduction, advection, dispersion) and the role of groundwater flow in extending thermal plumes. It highlights potential positive or negative interactions between installations and stresses the need for careful planning. Strategies such as spatial optimization, balanced thermal loads, and the use of geothermal databases are recommended. The chapter underlines that proper management of subsurface resources is essential for sustainable and efficient geothermal exploitation.

Shallow geothermal installations use the subsurface as a heat source (during building heating season) or as a heat sink (during the cooling season). As the thermal cycles start, the soil becomes thermally influenced by the installation in an area whose extension depends on a number of factors. Generally, closed loop systems, and hence EGs, show a smaller affected area compared to open loop systems (Barla et al., 2025). In urban environment, the relevant number of underground structures and infrastructures and the growing number of thermal users make the management of the subsurface an increasingly relevant planning issue (Baralis et al., 2020; Barla et al., 2020b). When a new EG is built in an urban environment, it will interact with the subsoil original conditions and previous installations, either with positive or negative drawbacks. Under certain circumstances different users may benefit from diverse thermal needs. Also, environmental effects may in these cases result reduced. It is the scope of this Chapter to provide information and guidance to tackle these aspects.

M. Barla et al., *Recommendations for the Conceptualisation, Design and Construction of Energy Geostructures*, SpringerBriefs in Applied Sciences and Technology,
https://doi.org/10.1007/978-3-032-30927-3_5

5.1 Reason of Thermal Interference

Thermal interference may happen whenever the heat from one installation is transported to the influence area of another installation. Thus, focus should be put on heat transfer mechanisms which includes conduction and advection with dispersion.

In presence of groundwater fluxes, apparent thermal conductivity results to be increased because of the advection and thus should be accounted while modelling (Ferguson, 2015).

Taking as an example the case of BHE design (which could also apply for an EG), heat advection and dispersion (Fig. 5.1) should be accounted (Alcaraz et al., 2016) because it reduces the time to reach steady state and affects the possibility of thermal storage.

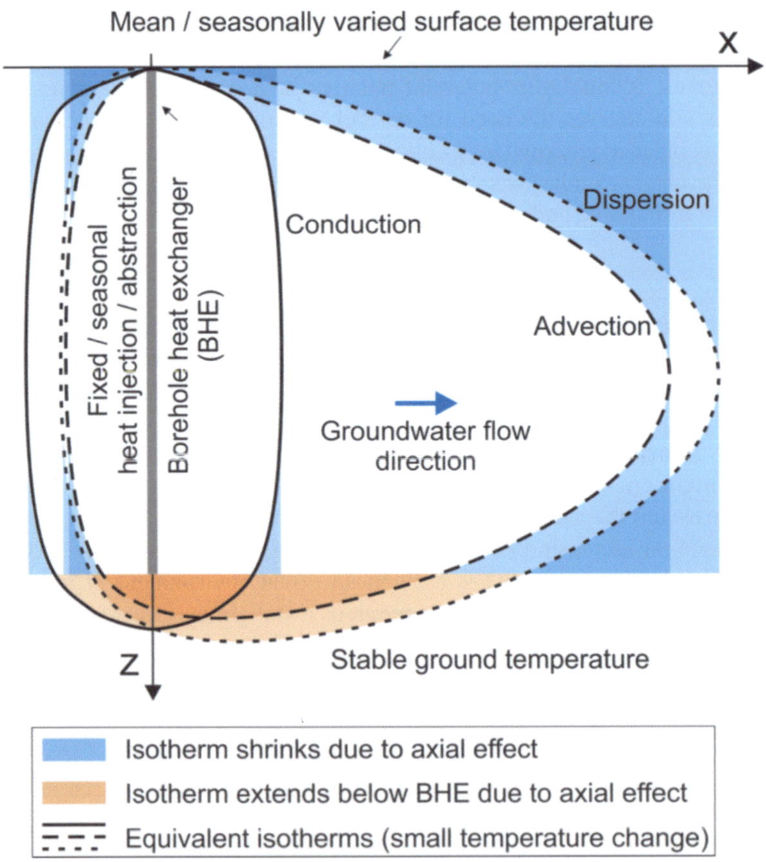

Fig. 5.1 Processes of heat transfer from BHE installed in homogeneous aquifer showing the effect of axial (in z direction) conduction and dispersion, groundwater advection and longitudinal and transverse dispersion. Equivalent Isotherms are shown for different mechanisms at same time (Pedchenko, 2019)

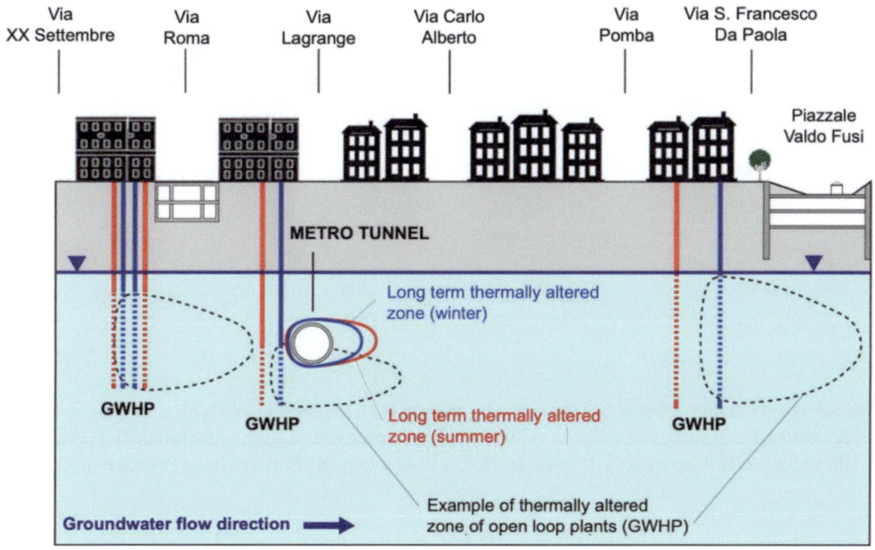

Fig. 5.2 Example illustration showing possible interferences in a specific urban context (courtesy of Geosolving Srl, Italy)

In an urban context, several structures and infrastructures may influence the thermal regime of the subsurface (Fig. 5.2). Thermal and hydraulic interferences may likely arise especially where higher concentration of anthropogenic structures are present.

In case of vertically linear structures (namely single/small number of energy piles, but also BHE), the effects that can be described in the surroundings are limited to the thermal ones. These can represent a source of interference in case the thermal affection interests the location of other EGs or geothermal installations. This occurs when the thermal plume extension intercepts the area where the installation is located. Even slow groundwater flows can significantly extend the thermal plume (Fig. 5.3). Groundwater flow almost always extends the plume in the direction of flow, except for high groundwater flow velocity. In this case the plume shrinks because groundwater advection decreases temperature at BHE wall and alters temperatures in the downstream area. Axial conduction for high groundwater flow is negligible.

Usually, BHE and energy piles are gathered in groups to provide sufficient energy to the users. On the one hand pile groups show a different behaviour from single installations, especially when spacing is low. On the other hand, the interaction among elements is explanatory of interference mechanisms. Multiple elements positioned along the groundwater direction accumulate thermal impact towards the downstream ones. In these conditions the efficiency of the upstream elements benefit from the groundwater advection the most (Fig. 5.4). High groundwater flow cannot efficiently reduce the longitudinal thermal plume extent, because of the superposition of the

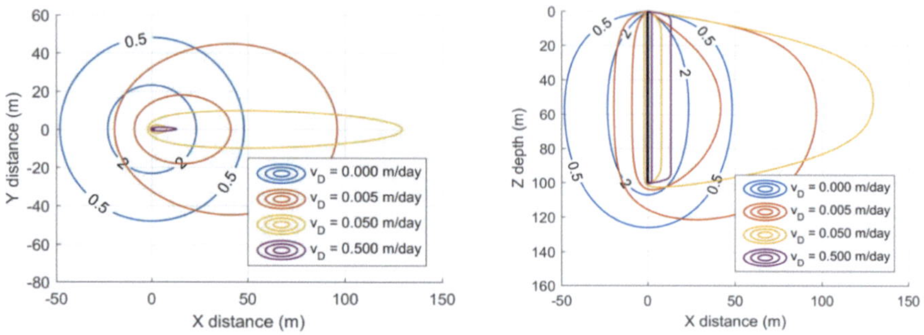

(a) (b)

Fig. 5.3 Isotherm for 0.5 and 2 K temperature change for a range of Darcy's groundwater velocities (v_D) generated using analytical solution (moving finite line source). Vertical borehole heat exchanger of 100 m length is located at X Y = [0, 0] m. (a) Plan view; (b) Profile view (Pedchenko, 2019)

thermal impacts from each individual element. Understanding of local hydrogeological conditions can therefore help to identify which installations will be less efficient and to optimise the design of the field accordingly.

Whenever a bi- or tridimensional geostructure is involved, interference may arise from the hydrogeological point of view too. Installations like energy walls, cut & cover energy tunnels and, to a lesser extent, bored energy tunnels perturb the hydraulic regime (dam effect). This is common to non-thermoactive structures too. The hydraulic regime alteration may lead to significant variation in the advective components of the heat transport with negative, but also beneficial, effects (Fig. 5.5).

5.2 Managing the Thermal Interference

This paragraph describes how to manage the thermal interference between EGs and neighbouring geothermal installations. Proper spatial planning allows to:

- optimize the design of the geostructure (e.g. operation, geometry, technology) to cause minimum conflicts,
- incorporate the potential interference at the feasibility stage,
- use the city subsurface (including the geothermal reservoir) in most efficient way, eliminate the "first come – first served" rule and mitigate the subsurface urban heat island (SUHI).

The design of an EG cannot be decoupled from the site-specific conditions. Among them thermal, hydraulic and mechanical interference with pre-existing other thermal or non-thermal groundwater usage are of particular importance. Possible interferences with underground structures have to be assessed as well. Nonetheless the thermal alteration induced on the groundwater body has to be quantified.

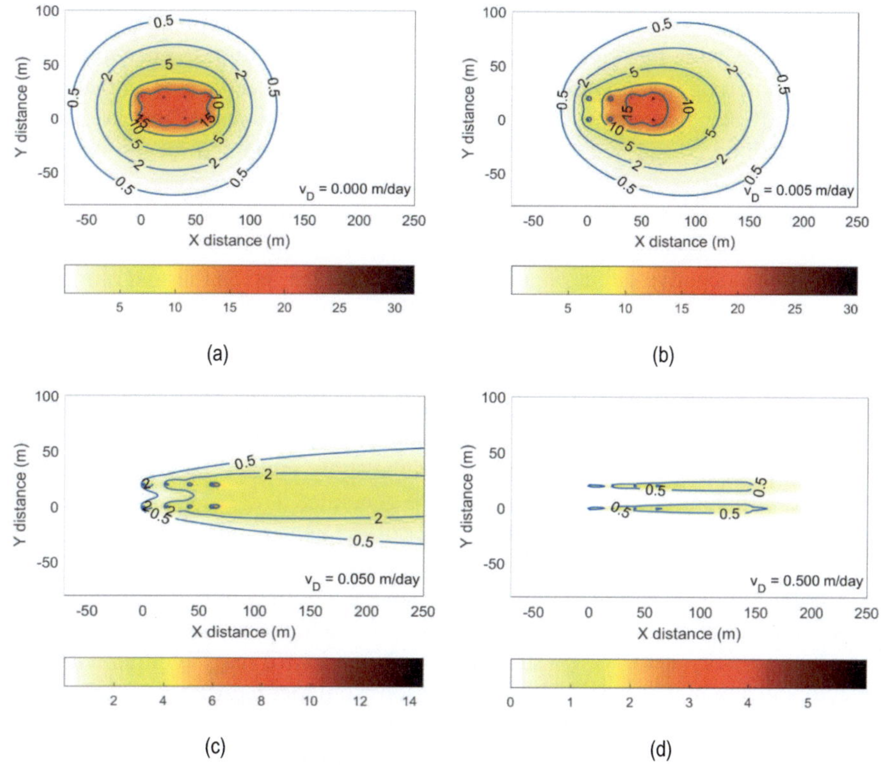

Fig. 5.4 Plan view of BHE field with isotherms of temperature change (K) for different Darcy groundwater velocities (v_D): (a) $v_D = 0$ m/day, (b)$v_D = 0.005$ m/day, (c) $v_D = 0.05$ m/day, (d) $v_D = 0.5$ m/day. Note that colour scales end with different maximal temperatures. Moving finite line source analytical solution (MFLS) was used (Pedchenko, 2019)

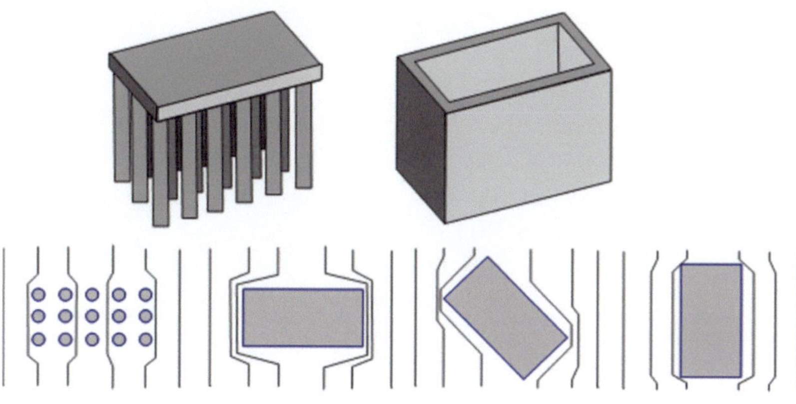

Fig. 5.5 Dam effect related to the type of infrastructure (CFMS and SYNTEC, 2017)

That is why, since the feasibility study phase of an EG, a census of other geothermal installations existing in a significant area around the site (with reference to both the closed-loop and open-loop circuits), should be carried out, especially in the case of high groundwater flow rates. As a first approximation, the size of the area subject to census can be considered equivalent to the extension adopted for the numerical models.

The interference may either involve the geostructure under design or be originated by it. From the thermal point of view the presence of an interference might result in a positive or a negative effect on a single installation. In fact, a thermal plume may be regarded either as a potential anthropogenic geothermal resource or as pollution depending on the downstream usage.

In case an aquifer is present the study should include upstream installations that may involve the site in their thermally altered zone (TAZ). Special attention should be devoted to open loop schemes which generally lead to wider thermal alterations (Fig. 5.6). Furthermore, the area included in the study in the downstream direction should be sufficiently extended to include the generated TAZ.

The study of the thermo-hydraulic impact of the geostructure should be used to optimize geometry, operation cycles and technology of the installation.

In order to reduce the depletion of the resource, strongly unbalanced heat loads should be avoided. In case of dominant heating season thermal load and low groundwater flow velocity, heat storage strategies can be evaluated (e.g. coupling with solar collectors in summer season). The optimization would let reduce the extension and magnitude of the TAZ, overcoming the "first come – first served" rule that regrettably governs shallow geothermal exploitation in many cities.

It is obvious that the optimization process has to balance two opposite needs: cost-efficiency of the installation and the reduction of environmental impact. It is therefore needed an iterative approach to reach the optimum balance. In addition

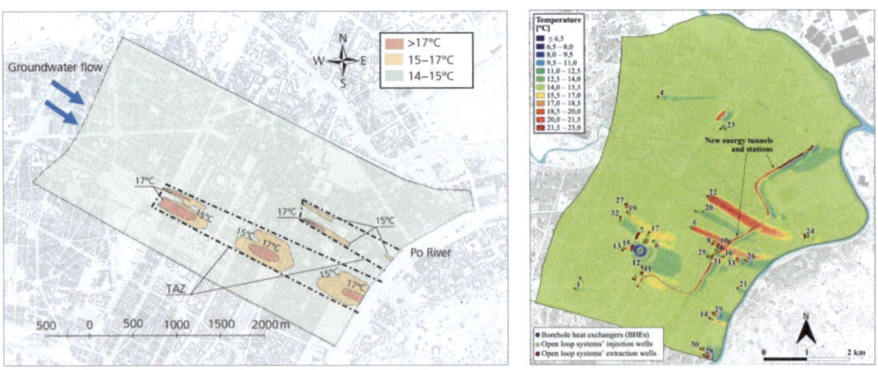

(a) (b)

Fig. 5.6 Examples of Thermally Altered Zone (TAZ) in an urban area due to geothermal exploitation from open loop systems (a, Barla et al., 2020b) and their impact on an energy tunnel (b, Barla et al. 2025) to this, a collaborative approach between stakeholders and authorities is definitely needed.

The designers can seek assistance from existing and available georeferred databases of geothermal installation. Such databases, which are constantly updated in metropolitan areas, are created and becoming more and more available by public authorities and geological surveys. In some urban areas regional groundwater flow models are available. They can be used either by the stakeholders or the public authorities. The former can benefit from regional models to set up local detailed models and to define the area where to look for potential interference sources and/or targets. The latter can oversee the process and preliminarily assess areas of potential impact and thus possible optimal location for monitoring.

Chapter 6
Execution, Testing and Monitoring

Abstract This chapter provides practical guidance on implementing energy geostructures, from design and construction to operation. It emphasizes early integration of the technology into the design process and coordination between engineering disciplines. Key aspects include installation of heat exchanger pipes, material selection, hydraulic system design, and quality control measures such as pressure testing. The chapter also outlines best practices for construction details across different geostructures (piles, walls, tunnel linings). Monitoring plays a central role, involving measurement of temperature, flow rate, pressure, and structural response to ensure performance and enable optimization over time. Overall, the chapter highlights execution quality and monitoring as critical for long-term system efficiency.

6.1 Adoption of the Technology into the Design Process

The design of Energy Geostructures (EGs) is a multidisciplinary process that has interactions with many steps of the design procedure for a building, a structure or an infrastructure.

Approaching EGs from the initial stages of a project is crucial for providing an optimized solution that maximizes the potential of this technology. Conducting a feasibility study at the outset allows for the identification of the thermal activation potential of various structures and several thermal activation scenarios can be identified. Such study should also correlate these scenarios with the overall energy needs for the specific project, along with the corresponding return on investment.

After the prior assessment the design of the EGs can be taken to the next step by implementing it into the current design procedures. Compared with conventional design, EGs require additional steps: investigation for the determination of thermal properties (see Chap. 2), computation of energy potential and thermo-mechanical assessment (see Chaps. 3 and 4). Following this, specific requirements for different stages of execution should be provided in order to ensure the successful implementation of the technology. Coordination between the specialized teams developing the EGs detailed design and the team responsible for the heat pump and hydraulic

M. Barla et al., *Recommendations for the Conceptualisation, Design and Construction of Energy Geostructures*, SpringerBriefs in Applied Sciences and Technology,
https://doi.org/10.1007/978-3-032-30927-3_6

plant is essential both at the design and execution stage. Moreover, the installation of a monitoring system at the execution phase allows for further optimization of the energy efficiency of the system during its lifetime based on the output of the operational performance.

6.2 Execution of the Energy Geostructures

The geothermal circuit typically consists of a pipe generally connected to the reinforcement cages. The pipes used are usually made of high-density cross-linked polyethylene (for example PE-Xa), with an external diameter of about 20 mm and a wall thickness of 2–2.3 mm. These pipes are capable of withstanding high pressures and temperatures as well as corrosion, ensuring adequate durability. The pipes are composed of three layers:

- an inner layer of high-density polyethylene;
- an intermediate layer made of polymeric material;
- an outer layer consisting of an ethylene–vinyl alcohol (EVOH) barrier, which prevents oxygen permeability.

The coil forms a closed circuit within which a heat-transfer fluid circulates, consisting of water mixed with ethylene or propylene glycol (typically in proportions of about 30%), which acts as the medium for heat transfer.

In the construction of energy geostructures, the heat-exchange pipes are fixed to the steel reinforcement before the concrete is poured, or before the reinforcement itself is placed in the concrete. The sum of all elements inside the geostructures (reinforcement cage, heat exchanger pipes, spacers, …) must be limited to ensure a good flow of the concrete.

Where possible, the heat-exchange pipes should be kept inside the steel reinforcement rather than outside, in order to avoid damage during installation and reductions in concrete cover. However, if placing them outside the reinforcement is possible and more convenient, it is essential to adopt measures to prevent possible damage to the coils during installation.

The heat-exchange pipes are usually fixed manually using plastic cable ties of the type used for electrical cables or metallic collars, in order to keep the pipes in position. Therefore, during construction planning, the additional time required for fixing the heat-exchange pipes must be taken into account.

Fixing must comply with the correct distribution of the heat-exchange pipes established during the thermal design phase, in order to ensure the estimated heat-exchange efficiency.

In general, when selecting all the materials used, it must be considered that during the concrete curing phase the temperatures inside the geostructure can reach high values.

The ends of the heat-exchange pipe coils require particular attention:

- with regard to the lower part (or in any case the end inserted into the ground), it is advisable to keep the heat-exchange pipes within the steel reinforcement, without letting them protrude, in order to avoid possible damage during transport or installation;
- in the upper part, on the other hand, short lengths of pipe are usually provided for hydraulic connection to the horizontal manifolds and therefore protrude from the reinforcement: these must be protected with suitable coverings during installation of the reinforcement cage, paying particular attention during trimming operations and in the subsequent phases of construction of slabs or horizontal civil works, avoiding damage to the terminal parts of the pipes and keeping them accessible to realize the integrity tests.

The connection points between the coils embedded in the steel reinforcement and the horizontal connections represent a mechanically weak point that requires particular care during the fixing phase. Even when the structure is composed of several parts (for example, a pile with several sections of steel reinforcement), special attention must be paid to the connection points between one circuit and another. Alternatively, accepting a lower energy output, it is possible to activate only the upper part of the geostructure, fixing the heat-exchange pipes only to the first section of reinforcement.

It is also important to pay attention to the proper functioning aspects of a traditional thermo-hydraulic system, such as reducing the risk of air presence in the circuit, minimizing pressure losses, and so on.

Standard practice also that, in order to avoid damage to the pipes due to the pressure of the fresh concrete, the circuit is temporarily filled with water, closed, and pressurized before the concrete pour is carried out. Hydraulic tightness tests are then recommended to be performed several times on the circuit, at each subsequent handover stage on site.

Furthermore, circulation of the heat-transfer fluid within the heat-exchange pipes and the manifolds leading to the heat pump is ensured by a circulation pump. The overall hydraulic circuit, which also includes the circulation pump, must be properly sized, taking into account the final overall system, flow rates, diameters, lengths, and the corresponding distributed and localized pressure losses.

The need to install control valves to isolate individual circuits should also be evaluated, allowing these sections to be isolated from the rest of the circuit in the event of damage to the piping, as well as enabling venting of the pipes. Venting the circuit is of fundamental importance, since excess air can lead to a reduction in the flow rate of the heat-transfer fluid, with a consequent decrease in the amount of heat that can be exchanged by the entire system, which is directly proportional to this parameter.

Some specific additional recommendations are given below.

For energy piles (Fig. 6.1), it is often recommended to reduce as much as possible the vicinity between activated and non-activated piles in order to minimize the thermal interferences and the induced stresses in the supported structures. If the pile is not mechanically connected to the supported structures (temporary pile, rigid

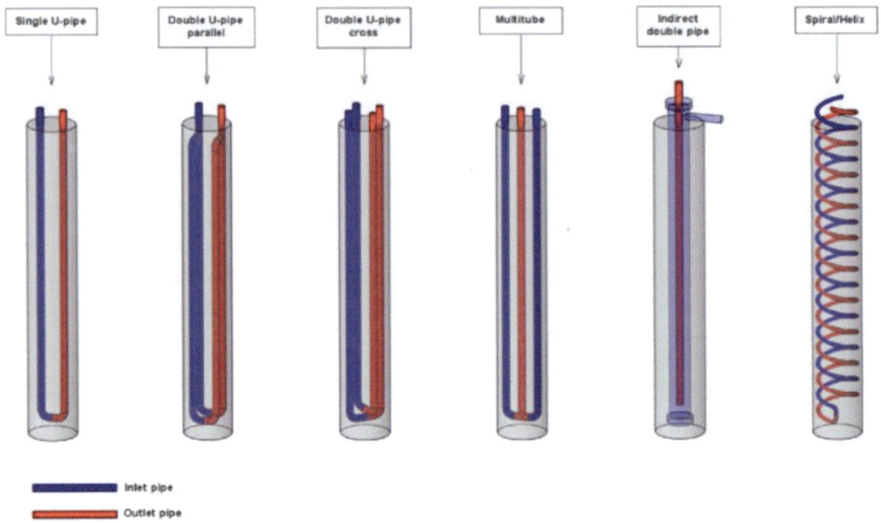

Fig. 6.1 Energy pile configurations (Fadejev et al., 2017)

inclusions) the only limits to consider are the ones imposed by the thermal design. In the case of vertical pipes, it implies that the pipes are attached to the transversal reinforcements and not the vertical ones.

Heat exchanger pipes should be at least 15 cm apart. For piles whose diameter is less than 40 cm, only one loop can be inserted. Then, for each 20 cm increase of the diameter, another loop can be inserted. For example, a pile with 60 cm diameter can be equipped with 2 loops and a pile with 80 cm diameter with 3 loops.

The bottom of the loop should be at least 25 cm above the pile/wall toe and the top of the loop should be at least 20 cm below the cut off level.

For slabs and shallow foundations, it is usually recommended to install the heat exchanger pipes under the geostructure itself, and, if needed, separated by an insulating material.

For diaphragm walls, the heat exchanger pipes are connected to the reinforcement cages and because of the phasing (one cage at once), every panel of the wall must have its own heat exchanger pipe.

If possible, the trimming of piles or diaphragm walls must be executed on fresh concrete. Otherwise, a specific protection of heat exchanger pipes is recommended. During the execution of top structures (pile heads, capping beam, …), the ends of the pipes must be placed in order not to be exposed to excessive loadings.

For mechanised excavated tunnels pipes are to be implemented in the concrete segments at the pre cast factory plant with similar precautions as for diaphragm walls. In case of conventional tunnelling, pipes are to be placed on top of the preliminary lining or behind it.

Many other geotechnical structures can also be equipped with heat exchanger pipes. No general rule can be given in these cases and specific studies must be carried out in order to assess their feasibility.

At every step of the project, the condition of tubes must be checked:

- when the pipes are delivered at the construction site, a visual check allows to determine their condition before attaching them to the reinforcement cages;
- once the pipes are attached on the reinforcement cages, they are pressurized (min. 4 bars) to check their integrity;
- the same process (pressurisation) applies once the geostructure is finished.

If the pressure test fails when the pipes are still accessible (e.g. before the concrete pouring in the case of cast-in-place pile or wall), the defective elements can still be replaced and the pressure test carried out again. Otherwise, the loop has to be considered impossible to activate and a spray can be used to detect the potential leakages around the head of the geostructure.

In any case, it is recommended to limit the number of energy geostructures connected in series in order to limit the impact of a defective loop. A common advice is to have no more than 5 piles in series or 2 diaphragm wall panels, however the proper number for a specific application has to be decided at the design stage to allow optimisation between hydraulic and energy aspects. A system of valves should allow easy separation of the different loops if one of them fails. It is also strongly recommended that a margin of at least 5% be allowed for the number of loops to obtain the supply of energy needs, even in case of defects on the tubes.

Once all the geostructures are realised, the heat exchanger pipes must be connected to the heating system. The pipes must lie on a material without sharp edges. A sand bed is generally used. The inlet and outlet pipes should be as far as possible from one another to limit thermal short circuits.

6.3 Monitoring

In the case of energy geostructures, monitoring may also include the assessment of thermal, hydraulic, and mechanical aspects induced by the thermal activation of the geostructures. The design of such a monitoring system should therefore provide for the definition of the significant quantities to be measured, the selection and placement of measuring instruments, and the methods for interpreting the experimental data. Among the useful quantities to be monitored are temperatures, flow rates, pressures, stresses, and deformations.

Appendix
Selected Examples

To provide the reader with relevant practical examples, three illustrative case studies are described in Figs. A.1, A.2 and A.3. They refer to shallow isolated foundations installed at the Oradea hospital in Romania, to energy diaphragm wall and slabs implemented at the Cleunay metro station in France and to energy tunnel linings adopted for the Turin metro line 1 in Italy. A brief description is provided for all cases together with key data including energy performance, technical characteristics, as well as monitoring and cost features.

M. Barla et al., *Recommendations for the Conceptualisation, Design and Construction of Energy Geostructures*, SpringerBriefs in Applied Sciences and Technology,
https://doi.org/10.1007/978-3-032-30927-3

ORADEA HOSPITAL

Shallow isolated foundations on energy piles

Location: **Oradea, Bihor County**
Country: **Romania**
Served building: **Hospital**
Year: **2018**
Typology: **Energy Pile Foundations**
Other energy sources: **Hybrid heat pump**
Reference: **Prodan et al. (2021) DOI:10.1007/978-3-030-64518-2_129**

DESCRIPTION OF THE PROJECT

This case study presents Romania's first building using isolated foundations on energy piles, implemented at the New Emergency Unit of the County Hospital in Oradea. Unlike conventional raft foundations, isolated footings were chosen for structural and economic reasons, supporting 230 energy piles that also serve as geothermal exchangers.

The system, embedded in stiff marl beneath compressive surface clays, supplies heating and cooling via a 150 kW ground source heat pump (GSHP) and a TABS system integrated into the building slabs. Thermo-mechanical monitoring included strain and temperature sensors in selected piles.

Numerical modeling showed significantly higher thermal-induced deformations in isolated foundations versus raft systems, due to reduced top rigidity and deeper thermal null points. These results emphasize the need for careful settlement control and specific design adjustments in isolated configurations.

The study confirms the viability of this system and provides a framework for broader adoption of energy geostructures in sustainable construction.

This first implementation advanced the application of energy geostructures in Romania. It offered a replicable model for sustainable, foundation-integrated heating and cooling systems in mid-rise institutional buildings, with insights for future design standards.

ENERGY INFORMATION

Average energy performance: **80 W/m of pile**
% of energy covered for the building:
100% for Heating
100% for Cooling

TECHNICAL DATA

Pile diameter: **0.6 m**
Pile length: **10.0 m**
Pipe layout: **spiral shape**
Ground condition: **altered clay marl**
Groundwater level: **-4 m**
Groundwater flow: **present**

MONITORING AND COST

Monitoring: **yes, sensors installed in 7 piles**
Payback period: **10 years**

Fig. A.1 Shallow isolated foundations for the Oradea hospital (Romania)

CLEUNAY METRO STATION

Diaphragm energy walls and slabs

Location: **Rennes**
Country: **France**
Served users: **Residential building**
Year: **2021**
Typology: **Energy walls and slab**
Other energy sources: **gas boiler**
Reference: **Sauvenier & Di Donna (2025)** **DOI:** 10.1016/j.enbuild.2025.116194

DESCRIPTION OF THE PROJECT

This case study describes energy walls and energy slabs installation at four stations of the metro line B in the city of Rennes in France. The construction of this metro line was part of a project designed to meet the challenges of sustainable development.

The new stations of Jules-Ferry, Saint-Anne, Saint-Germain and Cleunay have been equipped with a geothermal system embedded in the retaining walls and slabs, transforming them in energy geostructures.

The Cleunay station is equipped with 44 vertical pipes loops in the cast walls, with a total pipes length of 4056 m and a total equipped surface of 1532 m^2. The station slab is also equipped with a geothermal system made of 72 loops for a total pipes length of 5313 m and a total equipped surface of 1394 m^2.

The geothermal systems are active and properly working since 2021 (Cleunay), 2022 (St. Anne et St. Germain) and 2023 (Jules Ferry), exploiting between 23 to 32 W/m^2 of thermal energy. Jules-Ferry station is providing heating to 33 apartments in the Diell - Archipel Habitat building, Saint-Anne to 12 apartments, a commercial space and some offices in the Les Malouins - Archipel Habitat building, Saint-Germain to 10 apartments, one commercial and one sanitary space in Le Persan - Archipel Habitat building and Cleunay to 54 apartments in Emblem - Netoa building.

ENERGY INFORMATION

Winter extraction: **≈ 23-32 W/m^2**
Annual energy extracted: **53 MWh**
Summer injection: **none**
CO_2 saving: **9.3 tons/year**

TECHNICAL DATA

Bottom slab depth: **24 m**
Equipped walls surface: **1532 m^2**
Equipped slab surface: **1394 m^2**
Pipes material: **HDPE**
Pipes diameter: **25 mm**
Average flow rate: **≈ 9-12 m^3/h**
Ground condition: **schist & sandstone**

MONITORING AND COST

Monitoring: **yes**

Fig. A.2 Energy diaphragm wall and slabs for the Cleunay metro station (France)

TURIN METRO LINE 1

Energy tunnel prototype

Location: **Turin, Piedmont**
Country: **Italy**
Served building: **-**
Year: **2019**
Typology: **Energy Tunnels**
Other energy sources: **-**
Reference: **Barla et al. (2019) DOI: 10.1016/j.tust.2019.01.024; Insana & Barla (2020) DOI: 10.1016/j.renene.2020.01.086**

DESCRIPTION OF THE PROJECT

During the construction works of the South Extension of the Turin Metro Line 1, two energy segmental lining rings (European Patent Number: 16834047.9) for a total of 12 segments, were installed and tested in the Lingotto-Bengasi tunnel section. Each ring is 1.4 m long and 30.0 cm thick, with an internal diameter of 6.88 m. These are conceived to exchange heat with both ground and air through a double circuit of pipes arranged on the extrados and the intrados of the lining, respectively. Both circuits can exchange heat simultaneously or can be partially activated to endorse one or the other side.

PE-Xa pipes with an external diameter of 20.0 mm and a thickness of 2.0 mm were adopted. These were tied to the steel cage of each segment by hand during the precasting process to have a spacing between them of 20.0÷30.0 cm. This allowed no additional operational stages during the construction on site, other than the connections between pipes of different segments. Soon after the realisation, the energy tunnel efficiency was tested.

The segmental lining performed well, recording average winter extraction and summer injection rates of 48 W/m^2 and 60 W/m^2, respectively. These results are mainly due to the massive groundwater flow of Turin subsoil, which moves perpendicularly to the tunnel axis. The tunnel lining was monitored, too. While exchanging heat, this revealed to experience temperature oscillations of ±13.0÷19.0°C with respect to the far-field ground temperature. These correspond to relatively low-stress variations of 1.0÷1.5 MPa.

ENERGY INFORMATION

Winter extraction: **48 W/m^2**
Annual energy extracted: **-- GJ**
Summer injection: **60 W/m^2**
CO_2 saving: **-- tons/year**

TECHNICAL DATA

Overburden: **16.6 m**
External diameter: **7.48 m**
Installation length: **2.8 m**
Pipe layout: **circumferential setup**
Heat carrier fluid flow: **0.4÷0.7 m^3/h**
Ground condition: **coarse-grained**
Groundwater level: **-12 m**
Groundwater flow: **present**

MONITORING AND COST

Monitoring: **yes**
Costs: **-- $**
Payback period: **10 y**

Fig. A.3 Energy tunnel lining for Turin metro line 1 (Italy)

Bibliography

Alberdi-Pagola M, Poulsen SE, Loveridge F, Madsen S, Jensen RL (2018) Comparing heat flow models for interpretation of precast quadratic pile heat exchanger thermal response tests. Energy 145:721–733. https://doi.org/10.1016/j.energy.2017.12.104

Alcaraz M, García-Gil A, Vázquez-Suñé E, Velasco V (2016) Advection and dispersion heat transport mechanisms in the quantification of shallow geothermal resources and associated environmental impacts. Sci Total Environ 543(Part A):536–546. https://doi.org/10.1016/j.scitotenv.2015.11.022

Aresti L, Christodoulides P, Florides G (2018) A review of the design aspects of ground heat exchangers. Renew Sustain Energy Rev 92:757–773. https://doi.org/10.1016/j.rser.2018.04.053

Associazione Geotecnica Italiana (2025). Progettazione e realizzazione di geostrutture energetiche – Raccomandazioni, 108 pp.

Austin WA III, Yavuzturk C, Spitler JD (2000) Development of an in-situ system and analysis procedure for measuring ground thermal properties. ASHRAE Trans 106(1):365–379

Baralis M, Barla M (2024) ROGER: a method for determining the geothermal potential in urban areas. Geothermics 124(3):103148. https://doi.org/10.1016/j.geothermics.2024.103148

Baralis M, Barla M, Bogusz W, Di Donna A, Ryżyński G, Żeruń M (2020) Geothermal potential of the NE extension Warsaw (Poland) metro tunnels. Environ Geotech, 282–294 https://doi.org/10.1680/jenge.18.00042

Barla M, Di Donna A, Insana A (2019) A novel real-scale experimental prototype of energy tunnel. Tunn Undergr Space Technol 87:1–14. https://doi.org/10.1016/j.tust.2019.01.024

Barla M, Di Donna A, Santi A (2020a) Energy and mechanical aspects on the thermal activation of diaphragm walls for heating and cooling. Renewable Energy 147(2):2654–2663

Barla M, Di Donna A, Baralis M (2020b) City-scale analysis of subsoil thermal conditions due to geothermal exploitation. Environ Geotech 7(4):306–316. https://doi.org/10.1680/jenge.17.00087

Barla M, Insana A, Alvi MR (2025) Assessing the interaction of an energy tunnel with the underground thermal conditions in an urban area. Geothermics 130:103350

Barla M, Di Donna A (2018) Energy tunnels: concept and design aspects. Undergr Space 3(4):268–276, ISSN 2467-9674. https://doi.org/10.1016/j.undsp.2018.03.003

Bernier MA, Chahla A, Pinel P (2008) Long-term ground-temperature changes in geo-exchange systems. ASHRAE Trans 114(2):342–351

Bidarmaghz A, Makasis N, Fei W, Narsilio GA (2021) An efficient and sustainable approach for cooling underground substations. Tunn Undergr Space Technol 113:103986

Blum P, Campillo G, Kölbel T (2011) Techno-economic and spatial analysis of vertical ground source heat pump systems in Germany. Energy 36(5):3002–3011. https://doi.org/10.1016/j.energy.2011.02.044

M. Barla et al., *Recommendations for the Conceptualisation, Design and Construction of Energy Geostructures*, SpringerBriefs in Applied Sciences and Technology,
https://doi.org/10.1007/978-3-032-30927-3

Bourne-Webb PJ, Amatya B, Soga K, Amis T, Davidson C, Payne P (2009) Energy pile test at Lambeth College, London: geotechnical and thermodynamic aspects of pile response to heat cycles. Géotechnique 59(3):237–248. https://doi.org/10.1680/geot.2009.59.3.237

Bourne-Webb PJ, Bodas Freitas TM, da Costa Gonçalves RA (2016a) Thermal and mechanical aspects of the response of embedded retaining walls used as shallow geothermal heat exchangers. Energy Build 125:130–141

Bourne-Webb P, Burlon S, Javed S, Kürten S, Loveridge F (2016b) Analysis and design methods for energy geostructures. Renew Sustain Energy Rev 65:402–419. https://doi.org/10.1016/j.rser.2016.06.046

Brandl H (2006) Energy foundations and other thermo-active ground structures. Géotechnique 56(2):81–122. https://doi.org/10.1680/geot.2006.56.2.81

Burlon S, Habert J, Szymkiewicz F, Suryatriyastuti M, Mroueh H (2013) Towards a design approach of bearing capacity of thermo-active piles. In: European geothermal congress, Pisa, Italy

Carslaw HS, Jaeger JC (1959) Conduction of heat in solids, 2nd edn. Clarenden Press, Oxford

Cecinato F, Loveridge FA (2015) Influences on the thermal efficiency of energy piles. Energy 82:1021–1033. https://doi.org/10.1016/j.energy.2015.02.001

Cekerevac C, Laloui L (2004) Experimental study of thermal effects on the mechanical behaviour of a clay. Int J Numer Anal Meth Geomech 28(3):209–228. https://doi.org/10.1002/nag.332

Christodoulides P, Florides GA, Pouloupatis P (2016) A practical method for computing the thermal properties of a ground heat exchanger. Renewable Energy 94:81–89. https://doi.org/10.1016/j.renene.2016.03.035

Claesson J, Hellström G (2000) Analytical studies of the influence of regional groundwater flow by on the performance of borehole heat exchangers. In: Benner M (ed) Proceedings of the 8th international conference on thermal energy storage (Terrastock 2000), Stuttgart

Claesson J, Hellström G (2011) Multipole method to calculate borehole thermal resistances in a borehole heat exchanger. HVAC&R Res 17(6):895–911. https://doi.org/10.1080/10789669.2011.609927

Claesson J, Javed S (2011) An analytical method to calculate borehole fluid temperatures for time-scales from minutes to decades. ASHRAE Trans 117(2):279–288

Claesson J, Javed S (2020) Explicit multipole formula for the local thermal resistance in an energy pile: the line source approximation. Energies 13(20):5445. https://doi.org/10.3390/en13205445

Claesson J (2011) Radial heat flow for a pipe in a borehole in ground using Laplace solutions. Report on mathematical background. Technical Report, Department of Civil and Environmental Engineering, Chalmers University of Technology, Göteborg, Sweden

Comité Français de la Mécanique des Sols (CFMS), SYNTEC Ingénierie, SOFFONS-FNTP (2017) Recommandations pour la conception, le dimensionnement et la mise en œuvre des géostructures thermiques. Revue Française de Géotechnique 149:1–120. https://doi.org/10.1051/geotech/2017012

CSA/ANSI/IGSHPA C448:25 (2025) Design and installation of ground source heat pump systems for commercial and residential building. National Standard of Canada, American National Standard

Dalla Santa G, Galgaro A, Sassi R, Cultrera M, Scotton P, Müller J, Bertermann D, Mendrinos D, Pasquali R, Perego R, Pera S, Di Sipio E, Cassiani G, De Carli M, Bernardi A (2020) An updated ground thermal properties database for GSHP applications. Geothermics 85:101758. https://doi.org/10.1016/j.geothermics.2019.101758

Delage P (2013) On the thermal impact on the excavation damaged zone around deep radioactive waste disposal. J Rock Mech Geotech Eng 5(3):179–190. https://doi.org/10.1016/j.jrmge.2013.04.002

Di Donna A, Barla M (2016) The role of ground conditions on energy tunnels' heat exchange. Environ Geotech 3(4):214–224. https://doi.org/10.1680/jenge.15.00030

Di Donna A, Laloui L (2015) Response of soils subjected to thermal cyclic loading: experimental and constitutive study. Eng Geol 190:65–76. https://doi.org/10.1016/j.enggeo.2015.03.003

Di Donna A, Rotta Loria AF, Laloui L (2016) Numerical study of the response of a group of energy piles under different combinations of thermo-mechanical loads. Comput Geotech 72:126–142. https://doi.org/10.1016/j.compgeo.2015.11.010

Di Donna A, Barla M, Amis T (2017a) Energy geostructures: a collection of data from real applications. In: 15th international conference of the international association for computer methods and advances in geomechanics (15th IACMAG), Wuhan (China) 19–23 Ottobre 2017

Di Donna A, Cecinato F, Loveridge F, Barla M (2017b) Energy performance of diaphragm walls used as heat exchangers. Proc Inst Civ Eng-Geotech Eng 170(3):232–245

Di Donna A, Loveridge F, Piemontese M, Barla M (2021) The role of ground conditions on the heat exchange potential of energy walls. Geomech Energy Environ 25:100199. https://doi.org/10.1016/j.gete.2020.100199

Diao N, Li Q, Fang Z (2004) An analytical solution of the temperature response in geothermal heat exchangers with groundwater advection. Int J Therm Sci 43(10):987–995

Ditlefsen C, Sørensen I, Slott M, Hansen M (2014) Estimating thermal conductivity from lithological descriptions: A new web-based tool for planning of ground-source heating and cooling. Geol Surv Den Greenl Bull 31:55–58. https://doi.org/10.34194/geusb.v31.4660

Dong Y, McCartney JS, Lu N (2015) Critical review of thermal conductivity models for unsaturated soils. Geotech Geol Eng 33:207–221. https://doi.org/10.1007/s10706-015-9843-2

Dornberger SC, Rotta Loria A, Zhang M, Bu L, Epard J-L, Turberg P (2022) Heat exchange potential of energy tunnels for different internal airflow characteristics. Geomech Energy Environ 30:100229, ISSN 2352-3808. https://doi.org/10.1016/j.gete.2020.100229

Dupray F, Laloui L, Kazangba A (2014) Numerical analysis of seasonal heat storage in an energy pile foundation. Comput Geotech 55:67–77. https://doi.org/10.1016/j.compgeo.2013.08.004

EN 1997-1 (2004) Eurocode 7. Geotechnical design, general rules

Epting J, Böttcher F, Mueller MH, García-Gil A, Zosseder K, Huggenberger P (2020) City-scale solutions for the energy use of shallow urban subsurface resources—bridging the gap between theoretical and technical potentials. Renew Sustain Energy Rev 123:109768. https://doi.org/10.1016/j.renene.2019.09.021

Fadejev J, Simson R, Kurnitski J, Haghighat F (2017) A review on energy piles design, sizing and modelling. Energy 122:390–407. https://doi.org/10.1016/j.energy.2017.01.097

Farouki OT (1981) The thermal properties of soils in cold regions. Cold Reg Sci Technol 5(1):67–75. https://doi.org/10.1016/0165-232X(81)90041-0

Ferguson G (2015) Screening for heat transport by groundwater in closed geothermal systems. Groundwater 53(3):503–506. https://doi.org/10.1111/gwat.12162

Gawecka KA, Cui W, Taborda DMG, Potts DM, Zdravković L, Loukas A (2021) Predictive modelling of thermo-active tunnels in London Clay. Géotechnique 71(8):735–748. https://doi.org/10.1680/jgeot.19.P.376

Gehlin S (2002) Thermal response test—method development and evaluation. Ph.D. Thesis. Luleå University of Technology. 191 p.

GSHP Association (2012) Thermal pile design, installation and materials standards. 82 p.

Gupta A (2024) An analytical method to determine the amount of heat exchanged through thermally activated embedded retaining walls. Ph.D. Thesis. University of Leeds

Habert J, Burlon S (2015) Modelling thermo-active diaphragm walls. In: Second EAGE workshop on geomechanics and energy (Paper CP-466-00022). European Association of Geoscientists & Engineers. https://doi.org/10.3997/2214-4609.201414308

Hermans T, Nguyen F, Robert T, Revil A (2014) Geophysical methods for monitoring temperature changes in shallow low enthalpy geothermal systems. Energies 7(8):5083–5118. https://doi.org/10.3390/en7085083

Hillel D (2013) Introduction to soil physics, 1st edn. Academic Press. ISBN 978-0080918693

IGSHPA (2025) C448 Series-16 Design and installation of ground source heat pump systems for commercial and residential buildings

Ingersoll LR, Zobel OJ, Ingersoll AC (1954) Heat conduction with engineering, geological, and other applications, 3rd edn. New York, McGraw-Hill

Insana A, Barla M (2020) Experimental and numerical investigations on the energy performance of a thermo-active tunnel. Renewable Energy 152:781–792. https://doi.org/10.1016/j.renene.2020.01.086

Insana A, Barla M, Sulem J (2020) Energy tunnel linings thermo-mechanical performance: comparison between field observations and numerical modelling. E3S Web Conf 205:06008. https://doi.org/10.1051/e3sconf/202020506008

Insana A (2020) Thermal and structural performance of energy tunnels. Ph.D. Thesis. Politecnico di Torino. 436 p.

ISO 6946:2017 (2017) Building components and building elements—thermal resistance and thermal transmittance—calculation methods

Javed S, Fahlén P (2011) Thermal response testing of a multiple borehole ground heat exchanger. Int J Low-Carbon Technol 6(2):141–148. https://doi.org/10.1093/ijlct/ctr004

Jensen-Page L, Loveridge F, Narsilio GA (2019) Thermal response testing of large diameter energy piles. Energies 12:2700. https://doi.org/10.3390/en12142700

King W, Banks D, Findlay J (2012) Field determination of shallow soil thermal conductivity using a short-duration needle probe test. Q J Eng GeolHydrogeol 45:497–504. https://doi.org/10.1144/qjegh2012-00

Knellwolf C, Peron H, Laloui L (2011) Geotechnical analysis of heat exchanger piles. J Geotech Geoenviron Eng 137(10):890–902

Kürten S, Mottaghy D, Ziegler M (2015) A new model for the description of the heat transfer for plane thermo-active geotechnical systems based on thermal resistances. Acta Geotech 10(2):219–229. https://doi.org/10.1007/s11440-014-0311-6

Laloui L, François B (2009) ACMEG-T: soil thermoplasticity model. J Eng Mech 135(9):932–944. https://doi.org/10.1061/(ASCE)EM.1943-7889.0000011

Laloui L, Nuth M, Vulliet L (2006) Experimental and numerical investigations of the behaviour of a heat exchanger pile. Int J Numer Anal Meth Geomech 30(8):763–781. https://doi.org/10.1002/nag.499

Laloui L, Dupray F, Li C (2014) Heat-exchanger piles for the de-icing of bridges. Acta Geotech 9(3):413–423. https://doi.org/10.1007/s11440-014-0307-2

Laloui L, Di Donna A (2011) Understanding the thermo-mechanical behaviour of energy geostructures. Proc Inst Civ Eng—Civ Eng 164(CE4):184–191

Lamarche L, Beauchamp B (2007) A new contribution to the finite line-source model for geothermal boreholes. Energy Build 39:188–198

Lee S, Kim J, Park H (2021) Energy slabs: a review of thermal performance and influential factors. Renew Sustain Energy Rev 135:110253

Li Z, Wang C, Fu Q (2022) Effects of groundwater flow on thermal response test of deep borehole heat exchanger. Geothermics 106:102570. https://doi.org/10.1016/j.geothermics.2022.102570

Liu X, Spitler JD, Qu M, Shi L (2021) Recent developments in the design of vertical borehole ground heat exchangers for cost reduction and thermal energy storage. ASME J Energy Resour Technol 143(10):100803. https://doi.org/10.1115/1.4050418

Look B (2007) Handbook of geotechnical investigation and design tables. CRC Press

Loveridge F, Powrie W (2013a) Temperature response functions (G-functions) for single pile heat exchangers. Energy 57(C):554–564. https://doi.org/10.1016/j.energy.2013.04.060.

Loveridge F, Powrie W (2013b) Pile heat exchangers: thermal behaviour and interactions. Proc Inst Civ Eng—Geotech Eng 166(2):178–196. https://doi.org/10.1680/geng.11.00042

Loveridge F, Low J, Powrie W (2017) Site investigation for energy geostructures. Q J Eng GeolHydrogeol 50:158–168. https://doi.org/10.1144/qjegh2016-027

Loveridge F, Mrouch H, Vieira A (2020a) Editorial: shallow geothermal energy for buildings and infrastructure. Environ Geotech 7(4):223–225. https://doi.org/10.1680/jenge.2020.7.4.223

Loveridge F, McCartney JS, Narsilio GA, Sanchez M (2020b) Energy geostructures: a review of analysis approaches, in situ testing and model scale experiments. Geomech Energy Environ 22:100173. https://doi.org/10.1016/j.gete.2019.100173

Loveridge F, Woodman N, Javed S, Claesson J (2022) A fast approximate method for simulating thermal pile heat exchangers. Geomech Energy Environ 32:100368. https://doi.org/10.1016/j.gete.2022.100368

Lund J, Toth A (2020) Direct utilization of geothermal energy 2020 worldwide review. Geothermics 90:101915. https://doi.org/10.1016/j.geothermics.2020.101915

Ma C, Di Donna A, Dias D (2022) Numerical study on the thermo-hydro-mechanical behaviour of an energy tunnel in a coarse soil. Comput Geotech 151:105003, ISSN 0266-352X. https://doi.org/10.1016/j.compgeo.2022.105003

Ma C, Di Donna A, Dias D, Zhang J (2021) Numerical investigations of the tunnel environment effect on the performance of energy tunnels. Renewable Energy 172:1279–1292

Magraner T, Montero Á, Cazorla-Marín A, Montagud-Montalvá C, Martos J (2021) Thermal response test analysis for U-pipe vertical borehole heat exchangers under groundwater flow conditions. Renewable Energy 165(Part 1):391–404. https://doi.org/10.1016/j.renene.2020.11.029

Makasis N, Narsilio GA, Bidarmaghz A, Johnston IW, Zhong Y (2020) The importance of boundary conditions on the modelling of energy retaining walls. Comput Geotech 120:103399. https://doi.org/10.1016/j.compgeo.2019.103399

Man Y, Yang H, Diao N, Liu J, Fang Z (2010) A new model and analytical solutions for borehole and pile ground heat exchangers. Int J Heat Mass Transf 53(13–14):2593–2601

Maragna C, Loveridge F (2019) A resistive-capacitive model of pile heat exchangers with an application to thermal response tests interpretation. Renewable Energy 138:891–910, ISSN 0960-1481. https://doi.org/10.1016/j.renene.2019.02.012

Ma X, Grabe J (2010) Field test of a geothermal system in HafenCity Hamburg. W: Proceedings of GEOSHANGHAI, ASCE, Geotechnical Special Publication No. 204—Geoenvironmental Engineering and Geotechnics, 159–166

Mitchell JK (1993) Fundamentals of soil behavior, 2nd edn. Wiley

Modaressi H, Laloui L (1997) A thermo-viscoplastic constitutive model for clays. Int J Numer Anal Methods Geomech 21(5):313–335. https://doi.org/10.1002/(SICI)1096-9853(199705)21:5%3c313::AID-NAG872%3e3.0.CO;2-5

Mohamad Z, Fardoun F, Meftah F (2021) A review on energy piles design, evaluation, and optimization. J Clean Prod 292:125802. https://doi.org/10.1016/j.jclepro.2021.125802

Nicholson DP, Chen Q, Pillai A, Chendorain M (2013) Developments in thermal piles and thermal tunnel lining for city scale GSHP systems. In: Proceedings of the 38th workshop on geothermal reservoir engineering, pp 1–8

Nicholson DP, Chen Q, de Silva M, Winter A, Winterling R (2014) The design of thermal tunnel energy segments for Crossrail, UK. Proc Inst Civ Eng-Eng Sustain 167(3):118–134

Oladunjoye MA, Sanuade OA (2012) Thermal diffusivity, thermal effusivity and specific heat of soils in Olorunsogo Powerplant, southwestern Nigeria. Int J Res Rev Appl Sci 13(2):502–521. https://doi.org/10.5402/2012/591450

Pedchenko O (2019) The Effect of Fractures in Aquifers on the Performance of Borehole Ground Heat Exchangers, PhD Thesis, University of Southampton, UK

Perić D, Cossel AE, Sarna SA (2020) Analytical solutions for thermomechanical soil structure interaction in end-bearing energy piles. J Geotech Geoenvironmental Eng 146. https://doi.org/10.1061/(ASCE)GT.1943-5606.0002269

Poulos HG (1989) Pile behaviour—theory and application. Géotechnique 39(3):365–415. https://doi.org/10.1680/geot.1989.39.3.365

Poulsen SE, Andersen TR, Tordrup KW (2022) Full-scale demonstration of combined ground source heating and sustainable urban drainage in roadbeds. Energies 15(12). https://doi.org/10.3390/en15124505

Ramstad RK, Midttømme K, Liebel HT, Frengstad BS, Willemoes-Wissing B (2015) Thermal conductivity map of the Oslo region based on thermal diffusivity measurements of rock core samples. Bull Eng Geol Env 74(4):1275–1286. https://doi.org/10.1007/s10064-014-0701-x

Rees SW, Adjali MH, Zhou Z, Davies M, Thomas HR (2000) Ground heat transfer effects on the thermal performance of earth-contact structures. Renew Sustain Energy Rev 4(3):213–265. https://doi.org/10.1016/S1364-0321(99)00018-0

REN21 (2022) Renewables 2022 global status report. 308 pp. ISBN 978-3-948393-04-5

Rotta Loria AF, Di Donna A, Zhang M (2022) Stresses and deformations induced by geothermal operations of energy tunnels. Tunn Undergr Space Technol 124:104438. https://doi.org/10.1016/j.tust.2022.104438

Rotta Loria AF, Laloui L (2016) The interaction factor method for energy pile groups. Comput Geotech 80:121–137. https://doi.org/10.1016/j.compgeo.2016.07.002

Rotta Loria AF, Laloui L (2017) Impact of thermally induced soil deformation on the serviceability of energy pile groups. In: Ferrari A, Laloui L (eds) Advances in laboratory testing and modelling of soils and shales (ATMSS). ATMSS 2017. Springer Series in Geomechanics and Geoengineering. Springer, Cham. https://doi.org/10.1007/978-3-319-52773-4_50

Rotta Loria AF, Zannin J, Llabjani Q, Laloui L (2020) Analytical solution for describing the thermo-mechanical behavior of plane energy geostructures. E3S Web Conf 205:06009. https://doi.org/10.1051/e3sconf/2020205060099

Saggu R, Chakraborty T (2015) Cyclic thermo-mechanical analysis of energy piles in sand. Geotech Geol Eng 33(2):321–342. https://doi.org/10.1007/s10706-014-9798-8

Salciarini D, Gerola M, De Feudis S, Lupattelli A, Dalla Santa G, Capati G, Rafai M, Tihana J, Scerbo M, Di Donna A, Ravera E, Nicolino M, Insana A, Mroueh H, Barla M (2026) Database-driven analysis of global energy geostructures: diffusion, efficiency, and environmental performance. Renewable Energy 256(Part F):124373

Sanner B, Mands E, Sauer M, Grundmann E (2007) Technology, development status, and routine application of thermal response test. In: Proceedings of the European geothermal congress 2007, Unterhaching, Germany, 1–6

Scarlat N, Prussi M, Padella M (2022) Quantification of the carbon intensity of electricity produced and used in Europe. Appl Energy 305. https://doi.org/10.1016/j.apenergy.2021.117901

Shafagh I, Rees S, Loveridge F (2020) Investigations into thermal resistance of tunnel lining heat exchangers. E3S Web Conf 205:06006. https://doi.org/10.1051/e3sconf/202020506006

Shafagh I, Rees S, Mardaras IU, Janó MC, Carbayo MP (2020b) A model of a diaphragm wall ground heat exchanger. Energies 13(2):300. https://doi.org/10.3390/en13020300

Shafagh I, Rees SJ (2019). Analytical investigations into thermal resistance of diaphragm wall heat exchangers. In: Proceedings of the European geothermal congress 2019, 1–6

SIA (2005) D 0190 Utilisation de la chaleur du sol par des ouvrages de fondation et de soutènement en béton: guide pour la conception, la réalisation et la maintenance. Swiss Society of Engineers and Architects, Geneva, Switzerland

Soga K, Rui Y (2016) Energy geostructures. In: Rees SJ (ed) Advances in ground-source heat pump systems. Woodhead Publishing, pp 185–221

Soga K, Rui Y, Nicholson D (2015) Behaviour of a thermal wall installed in the Tottenham Court Road station box. In: Proceedings of the Crossrail conference, Crossrail Ltd and Federation of Piling Specialists, London, 112–119

Spitler JD, Bernier M (2016) 2—Vertical borehole ground heat exchanger design methods. In: Rees SJ (ed) Advances in ground-source heat pump systems. Woodhead Publishing, 29–61, ISBN 9780081003114. https://doi.org/10.1016/B978-0-08-100311-4.00002-9

Spitler JD, Gehlin SEA (2015) Thermal response testing for ground source heat pump systems—an historical review. Renew Sustain Energy Rev 50:1125–1137. https://doi.org/10.1016/j.rser.2015.05.061

Staffell I, Brett DJL, Brandon N, Hawkes A (2012) A review of domestic heat pumps. Energy Environ Sci 5:9291–9306. https://doi.org/10.1039/C2EE22653G

Sterpi D, Coletto A, Mauri L (2017) Investigation on the behaviour of a thermo-active diaphragm wall by thermo-mechanical analyses. Geomech Energy Environ 9(1):1–20. https://doi.org/10.1016/j.gete.2016.10.001

Sterpi D, Tomaselli G, Angelotti A (2020) Energy performance of ground heat exchangers embedded in diaphragm walls: field observations and optimization by numerical modelling. Renewable Energy 147:2748–2760. https://doi.org/10.1016/j.renene.2018.11.102

Sun M, Xia C, Zhang G (2013) Heat transfer model and design method for geothermal heat exchanger tubes in diaphragm walls. Energy Build 61:250–259. https://doi.org/10.1016/j.enbuild.2013.02.017

Suryatriyastuti ME (2013) Numerical study of the thermo-active piles behavior in cohesionless soils. Ph.D. Thesis. University of Lille. 176 p.

Verein Deutscher Ingenieure (2010) VDI 4640 Blatt 1: Thermische Nutzung des Untergrundes – Grundlagen, Genehmigungen, Umweltaspekte [Thermal use of the underground – Fundamentals, approvals, environmental aspects]. VDI Verlag, Düsseldorf

Vieira A, Alberdi-Pagola M, Barla M, Christodoulides P, Florides G, Insana A, Javed S, Maranha J, Milenic D, Prodan I, Salciarini D (2022) Site characterization for the design of thermoactive geostructures. Soils Rocks 45(1):e2022001022

Vieira A, Alberdi-Pagola M, Christodoulides P, Javed S, Loveridge F, Nguyen F, Cecinato F, Maranha J, Florides G, Prodan I, Van Lysebetten G, Ramalho E, Salciarini D, Georgiev A, Rosin-Paumier S, Popov R, Lenart S, Erbs Poulsen S, Radioti G (2017) Characterisation of ground thermal and thermo-mechanical behaviour for shallow geothermal energy applications. Energies 10(12):2044. https://doi.org/10.3390/en10122044

Wagner R, Clauser C (2005) Evaluating thermal response tests using parameter estimation for thermal conductivity and thermal capacity. J Geophys Eng 2(4):349–356. https://doi.org/10.1088/1742-2132/2/4/S08

Zannin J, Ferrari A, Kazerani T, Koliji A, Laloui L (2022) Experimental analysis of a thermoactive underground railway station. Geomech Energy Environ 29:100275. https://doi.org/10.1016/j.gete.2021.100275

Zhang G, Cao Z, Liu Y, Chen J (2021) Field test and numerical simulation on the long-term thermal response of PHC energy pile in layered foundation. Sensors 21:3873. https://doi.org/10.3390/s21113873

Zhang G, Xia C, Sun M, Zou Y, Xiao S (2013) A new model and analytical solution for the heat conduction of tunnel lining ground heat exchangers. Cold Reg Sci Technol 88:59–66. https://doi.org/10.1016/j.coldregions.2013.01.003